T0211902

Path Planning and Tracking for Vehicle Collision Avoidance in Lateral and Longitudinal Motion Directions

Synthesis Lectures on Advances in Automotive Technology

Editor
Amir Khajepour, *University of Waterloo*

The automotive industry has entered a transformational period that will see an unprecedented evolution in the technological capabilities of vehicles. Significant advances in new manufacturing techniques, low-cost sensors, high processing power, and ubiquitous real-time access to information mean that vehicles are rapidly changing and growing in complexity. These new technologies—including the inevitable evolution toward autonomous vehicles—will ultimately deliver substantial benefits to drivers, passengers, and the environment. Synthesis Lectures on Advances in Automotive Technology Series is intended to introduce such new transformational technologies in the automotive industry to its readers.

Narrow Tilting Vehicles: Mechanism, Dynamics, and Control
Chen Tang and Amir Khajepour
2019

Dynamic Stability and Control of Tripped and Untripped Vehicle Rollover
Zhilin Jin, Bin Li, and Jungxuan Li
2019

Real-Time Road Profile Identification and Monitoring: Theory and Application
Yechen Qin, Hong Wang, Yanjun Huang, and Xiaolin Tang
2018

Noise and Torsional Vibration Analysis of Hybrid Vehicles
Xiaolin Tang, Yanjun Huang, Hong Wang, and Yechen Qin
2018

Smart Charging and Anti-Idling Systems
Yanjun Huang, Soheil Mohagheghi Fard, Milad Khazraee, Hong Wang, and Amir Khajepour
2018

Design and Avanced Robust Chassis Dynamics Control for X-by-Wire Unmanned Ground Vehicle
Jun Ni, Jibin Hu, and Changle Xiang
2018

Electrification of Heavy-Duty Construction Vehicles
Hong Wang, Yanjun Huang, Amir Khajepour, and Chuan Hu
2017

Vehicle Suspension System Technology and Design
Avesta Goodarzi and Amir Khajepour
2017

Path Planning and Tracking for Vehicle Collision Avoidance in Lateral and Longitudinal Motion Directions

Jie Ji, Hong Wang, and Yue Ren

ISBN: 978-3-031-00379-0   paperback
ISBN: 978-3-031-01507-6   ebook
ISBN: 978-3-031-00011-9   hardcover

DOI 10.1007/978-3-031-01507-6

A Publication in the Springer series
SYNTHESIS LECTURES ON ADVANCES IN AUTOMOTIVE TECHNOLOGY

Lecture #12
Series Editor: Amir Khajepour, *University of Waterloo*
Series ISSN
Print 2576-8107   Electronic 2576-8131

Path Planning and Tracking for Vehicle Collision Avoidance in Lateral and Longitudinal Motion Directions

Jie Ji
College of Engineering and Technology, Southwest University, China

Hong Wang
School of Vehicle and Mobility, Tsinghua University, China

Yue Ren
College of Engineering and Technology, Southwest University, China

SYNTHESIS LECTURES ON ADVANCES IN AUTOMOTIVE TECHNOLOGY #12

ABSTRACT

In recent years, the control of Connected and Automated Vehicles (CAVs) has attracted strong attention for various automotive applications. One of the important features demanded of CAVs is collision avoidance, whether it is a stationary or a moving obstacle. Due to complex traffic conditions and various vehicle dynamics, the collision avoidance system should ensure that the vehicle can avoid collision with other vehicles or obstacles in longitudinal and lateral directions simultaneously. The longitudinal collision avoidance controller can avoid or mitigate vehicle collision accidents effectively via Forward Collision Warning (FCW), Brake Assist System (BAS), and Autonomous Emergency Braking (AEB), which has been commercially applied in many new vehicles launched by automobile enterprises. But in lateral motion direction, it is necessary to determine a flexible collision avoidance path in real time in case of detecting any obstacle. Then, a path-tracking algorithm is designed to assure that the vehicle will follow the predetermined path precisely, while guaranteeing certain comfort and vehicle stability over a wide range of velocities. In recent years, the rapid development of sensor, control, and communication technology has brought both possibilities and challenges to the improvement of vehicle collision avoidance capability, so collision avoidance system still needs to be further studied based on the emerging technologies.

In this book, we provide a comprehensive overview of the current collision avoidance strategies for traditional vehicles and CAVs. First, the book introduces some emergency path planning methods that can be applied in global route design and local path generation situations which are the most common scenarios in driving. A comparison is made in the path-planning problem in both timing and performance between the conventional algorithms and emergency methods. In addition, this book introduces and designs an up-to-date path-planning method based on artificial potential field methods for collision avoidance, and verifies the effectiveness of this method in complex road environment. Next, in order to accurately track the predetermined path for collision avoidance, traditional control methods, humanlike control strategies, and intelligent approaches are discussed to solve the path-tracking problem and ensure the vehicle successfully avoids the collisions. In addition, this book designs and applies robust control to solve the path-tracking problem and verify its tracking effect in different scenarios. Finally, this book introduces the basic principles and test methods of AEB system for collision avoidance of a single vehicle. Meanwhile, by taking advantage of data sharing between vehicles based on V2X (vehicle-to-vehicle or vehicle-to-infrastructure) communication, pile-up accidents in longitudinal direction are effectively avoided through cooperative motion control of multiple vehicles.

KEYWORDS

collision avoidance, connected and automated vehicles, path planning and tracking, artificial potential field, model predictive control, robust control, autonomous emergency braking, cooperative motion control, V2X communication

Contents

Acknowledgments

We regard this book as a way to express our appreciation and respect to Dr. Amir Khajepour who invited us to participate and write sections of the book, and put forward many constructive suggestions for the manuscript. I would also like to give my appreciation to the co-authors of this book, Dr. Hong Wang and Dr. Yue Ren—without their valuable contributions and helpful comments finishing this book would not have been possible. My sincere thanks also goes to all the members of GIVE laboratory at Southwest University for being great teammates on all our collaborative works. Finally, we are also thankful to Morgan & Claypool Publishers for providing the opportunity for writing this book, along with their consistent encouragement and support throughout this project.

Jie Ji
July 2020

CHAPTER 1

Introduction

1.1 BACKGROUND

Vehicle safety on the road is the kind of work that never stops!

Owing to the rapid increase in traffic density, vehicle safety has become a crucial issue in modern transportation systems. According to the World Health Organization, road traffic injuries are the leading cause of death for children and young adults aged 5–29 years and the 8th leading cause of death for all age groups. Road traffic injuries cause an estimated 1.35 million deaths worldwide and cost most countries 3% of their gross domestic product every year [1]. In China, the Chinese Ministry of Public Security data shows 63,194 people were killed and 258,532 people were injured in 244,937 motor vehicle crashes in 2018 (as shown in Fig. 1.1). That is, one person is killed every 8 minutes in China [2].

Vehicle safety is a complex problem and is influenced by a series of risk factors, e.g., drivers, vehicles, and traffic environment. Through the analysis of the causes of road traffic accidents, it has been found that driver error is a significant factor in accident causation, in fact, the European Accident Research and Safety Report 2013 showed that nearly 90% of road accidents were due to driver's inattention [3]. Therefore, reducing the number of accidents and fatalities on the roadways and highways which are caused by driver error has always been an important topic for research in automotive and transportation systems.

Many researchers believe connected and automated vehicles will provide a potential solution to the safety problems in future transportation, and various studies showed that nearly 40–50% of traffic accidents due to driver error could be reduced by installing Advanced Driver Assistance Systems (ADAS) in vehicles [4]. As consumers pay more and more attention to the active safety of new vehicles, ADAS (e.g., ACC, ABS, ESP, LKA, AEB, etc.) have been developed and became one of the principal priorities for most vehicle manufacturers, among them, Collision Avoidance Systems (CAS) fall into this category and have naturally designed to decrease the possibility and severity of traffic accidents on road during the last decade. The task of a CAS is to track objects of potential collision risk and determine necessary action to reduce the number of accidents and fatalities on the roadways and highways [5]. The CAS represents an opportunity to significantly mitigate the main influence factor of human error and increase transportation safety.

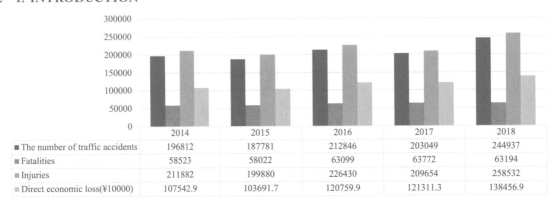

	2014	2015	2016	2017	2018
■ The number of traffic accidents	196812	187781	212846	203049	244937
■ Fatalities	58523	58022	63099	63772	63194
■ Injuries	211882	199880	226430	209654	258532
■ Direct economic loss(¥10000)	107542.9	103691.7	120759.9	121311.3	138456.9

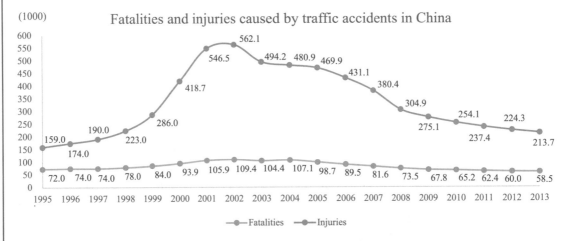

Figure 1.1: Statistics on traffic accidents in China by year, 1995–2018.

1.2 COLLISION AVOIDANCE SYSTEMS FOR A SINGLE VEHICLE

Avoiding collisions between vehicles and other obstacles is considered to be the most obvious way to improve the safety performance of intelligent transportation systems, and the related technologies have always been the subject of extensive research both in the fields of autonomous robotics and intelligent vehicles. Since research on autonomous robots has been going on for more than three decades, the lessons learned in this field should, to some extent, benefit the research on intelligent vehicles [6]. However, the collision avoidance method for autonomous robots generally treats the robot as a simplified point-mass motion model and computes a collision-free path in an arbitrary direction [7]. But for an intelligent vehicle, it is not difficult to appreciate that the maneuverability is a critical factor in lateral collision avoidance systems. With the increasing speed of vehicles, the risk of collision between vehicles and obstacles heightens,

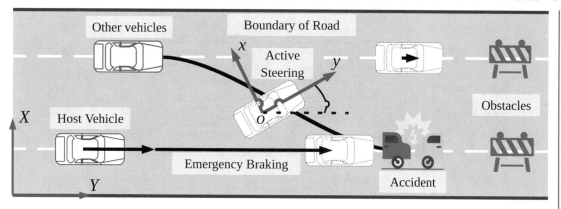

Figure 1.2: Lateral and longitudinal maneuvers for collision avoidance.

and the losses caused by collision are correspondingly more severe. Compared with the collision avoidance system for robots, the collision avoidance controller for intelligent vehicles face more complex road conditions and needs to consider more factors of kinematics and dynamics.

The objective of collision avoidance system is to design a vehicle control algorithm to avoid an imminent accident. Longitudinal control (i.e., emergency braking only), and lateral control (i.e., active steering only), are possible choices of actuation configuration for collision avoidance maneuver [8]. Figure 1.2 represents the aforementioned two maneuvers.

In most cases, the driver tries to avoid crash accidents by braking. In fact, the brake-based collision avoidance system is a better choice in the low speed range. Experimental studies and market feedback have evidenced the effectiveness of Forward Collision Warning (FCW) and Autonomous Emergency Braking (AEB) in improving driving safety, as well as the subjective evaluation of drivers [5]. However, in higher speed ranges, (e.g., over 60 km/h), longitudinal collision avoidance controllers are of limited benefit for preventing head-on collisions or rear-end collisions when there is insufficient separation between road vehicles. A detailed analysis of 635 accidents in GIDAS database showed that more than 80% of all drivers did not steer in a rear-end accident scenario. Further analysis showed that, in 24% of these accidents, steering would have been the better choice and the space conditions for an evasion maneuver had been fulfilled [9]. In these circumstances, if space is available in an adjacent lane, CAS based on evasion maneuvers are better suited, the maneuvers in this case can be completed in a shorter distance than that required to stop the vehicle.

Next, we will introduce and analyze the main collision avoidance technologies applied to connected and automated vehicles in the longitudinal and lateral motion directions, respectively.

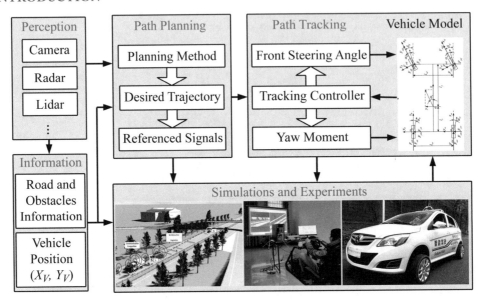

Figure 1.3: Hierarchical architecture of a collision avoidance system.

(1) Lateral Collision Avoidance Systems

In the lateral motion control studies, collision avoidance capability for vehicles can be incorporated into autonomous driving system through the integration of appropriate path-planning and tracking modules. With the information from onboard sensors about road ahead and other vehicles, a path-planning module can be constructed that avoids crashes and is feasible for the vehicle dynamics under the given traffic conditions. The objectives of collision avoidance should also be implemented by a path-tracking controller which can guarantee the vehicle's tracking capability to ensure that the vehicle is followed safely.

Under high speed driving condition, path-planning and tracking for collision avoidance in lateral direction is challenging for several reasons. The first challenge is caused by constraints on trajectory feasibility arising from the vehicle dynamics and physical limits such as actuator saturation or maximum friction forces. These constraints make it more difficult to design path-planning and tracking controllers. Another challenge is posed by the presence of unstable equilibrium points in the vehicle yaw dynamics depending on the vehicle speed, steering angle, and tire properties. Path-tracking must be designed to avoid these regions of instability in the dynamics. Finally, algorithm computation time must be short at high speed to allow a timely response to changes in the environment. From the above, it is still a challenging problem to design the collision avoidance algorithms for high-speed intelligent vehicles in lateral motion direction under vehicle dynamics and kinematics constraints.

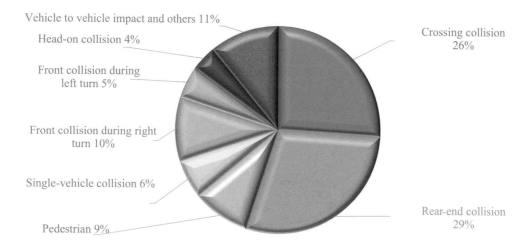

Figure 1.4: Ratio of accident type.

The development and application of lateral collision avoidance system for intelligent vehicles requires a variety of the-state-of-the-art technologies, such as sensing, control, decision-making, and other key technologies. According to different stages of collision avoidance, a hierarchical structure of collision avoidance system can be divided into four layers, namely environmental perception layer, decision-making layer, path-planning layer, and path-tracking layer, among which path-planning and path-tracking are the most basic and critical issues affecting the safety and comfort performance of collision avoidance system, as depicted in Fig. 1.3.

In this book, path-planning and path-tracking problems for collision avoidance will be emphasized, and we will introduce and compare different approaches to solve these two problems.

(2) Longitudinal Collision Avoidance Systems

In the longitudinal motion control studies, accidents analysis shows that rear-end crashes are the most frequently occurring type of collision. In fact, roughly 1.7 million rear-end collisions take place in the United States each year [10], accounting for 29% of all road traffic accidents, as shown in Fig. 1.4. These crashes result in a substantial number of injuries and fatalities each year, with about 1,700 people killed and another 500,000 are injured in the crashes, and it constitutes a significant portion of highway accidents.

There are two main reasons for the rear-end collision:

(a) a late reaction or no response from the driver (e.g., the driver approaches a traffic jam and is not attentive); or

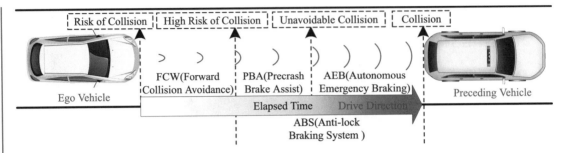

Figure 1.5: The longitudinal collision avoidance systems.

(b) insufficient headway (where the driver does not keep sufficient distance from the vehicle in front, so if the vehicle ahead suddenly stops, the driver does not have enough time to react).

At present, a number of automotive active safety systems are designed to reduce the possibility of collision in the direction of longitudinal motion. For example, Forward Collision Warning (FCW), Precrash Brake Assist (PBA), Autonomous Emergency Braking (AEB), and Antilock Braking System (ABS) are implemented in current and near-term passenger vehicles. At the last possible moment before a collision is likely to happen, the FCW system uses visual, auditory, and/or tactile means to warn the driver of an impending collision and buy sufficient time for the driver to take corrective action [11]. Whether on wet or dry roads, heavy application of the brakes on a vehicle may result in wheel lock and lead to accidents. An ABS prevents the wheels from locking up, which maintains the steering ability of vehicles while maximizing the braking force so that the stopping distance is reduced [12]. When making emergency stops, about half of all drivers do not press the brake fast enough or hard enough to make full use of their vehicle's braking power [13]. PBA is triggered when the vehicle recognizes an emergency-braking scenario and amplifies braking input when the driver applies the brake. Finally, AEB alerts a driver to an imminent crash and helps him use the maximum braking capacity of the vehicle [14]. The most basic AEB systems work at low speeds to prevent or reduce the severity of collisions by autonomously adding to the vehicle's braking deceleration, even if the situation becomes critical and no human response is made. More sophisticated systems work across a wider speed range and are triggered last, closest to the collision, as shown in Fig. 1.5.

Collision avoidance systems in longitudinal motion direction that can avoid rear-end crashes are a promising method of reducing collision-related injuries and property damage. All collision avoidance systems use sensors to detect obstacles ahead and assess whether a collision is likely. Today's most advanced sensors for collision avoidance system typically employ radar, LiDAR, camera, or a "stereo" system with two cameras. These systems can also use input from traditional sensors or interact with other systems such as speed sensors and steering angle sensors. The control unit will usually start by warning the driver that a collision is likely and that they need to brake, using dashboard warning lights or an audible alarm. If the driver fails to take action, the "autonomous" part of the system will kick in and apply the part or full brakes automatically. A significant benefit of the collision avoidance systems is their ability to protect vulnerable road users [15].

1.3   COOPERATIVE COLLISION AVOIDANCE SYSTEMS FOR MULTIPLE-VEHICLES

For a single vehicle, the collision avoidance technology in the longitudinal motion direction can effectively improve the active safety of the vehicle and reduce the incidence of rear-end collision. In a recent survey, the European Union (EU) pointed out that introducing the AEB system could reduce the annual number of deaths and serious injuries in vehicle accidents by more than 8,000 and 20,000, respectively [16]. However, the existing radar and camera-based systems can only detect those vehicles that are within the employed sensors' measurement ranges, and blind spots may occur owing to obstacles. In addition, under bad weather conditions, detection becomes impossible or the detection accuracy drops significantly. In order to overcome the limitations of sensor-based systems, recently, with the advancement of wireless communication technology, vehicle-to-vehicle (V2V) and vehicle-to-infrastructure (V2I) communications are setting the basis for establishing the cooperative collision avoidance technology by having vehicles interact with each other and with the surrounding infrastructure, sharing information about the environment and improving overall traffic awareness. Therefore, V2X communication based cooperative collision avoidance system have been the subject of intense research worldwide in government and industry consortia [17].

Take AEB, for example. The existing studies on AEB systems were conducted based on the performance of sensors employed in vehicles, but these sensors are very vulnerable to severe weather such as rain, snow, fog, and so on. Therefore, there is a need to improve the performance of AEB system that can solve the problem of rear-end collisions in varying weather conditions [18, 19]. To put the problem into context, cooperative collision avoidance (CCA) technologies display a vista for improving the traffic safety in all-weather conditions by deploying V2X communications [20]. The following vehicle can detect the sudden slow down or stop of the preceding vehicles according to the real-time position and speed information received by the wireless communication module, then CCA technologies provide drivers with more reaction time or automatically brake the vehicle to help avoid a collision. Another limitation of the AEB system, that is, the blind zone occurring at an intersection, can be partially solved by employing V2I communication in traffic lights at intersections [21]. With the help of the V2X communication technologies, vehicles equipped with AEB system can effectively detect traffic accidents beyond the visual range in advance, and improve the overall active safety of the transportation system in severe weather conditions.

Many studies and literature have confirmed the effectiveness of V2X communication in reducing the number of traffic accidents and improving traffic efficiency. In addition, the po-

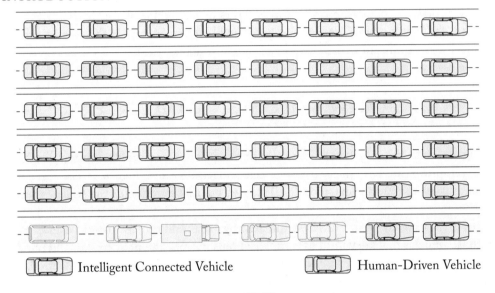

Figure 1.6: Mixed traffic flow with ICVs and HDVs.

tential and possibility of applying CCA technology to more traffic scenarios have been proposed [22]. However, when the CCA system is applied to the actual traffic environment, the following problems will still be encountered. First, the range, mode and topology of communication and package loss have influence on the performance of CCA systems. Moreover, the time delay in sensor detection (perception), signal transmission (communication), and driver/actuator response (action) will accumulate and affect the stability of the CCA systems. Second, the actual situation is that those equipped and unequipped vehicles would co-exist in general traffic flows for a long time. Therefore, it is necessary to study the vehicle collisions in platoons where only a fraction of vehicles are equipped with V2X communications [23], as shown in Fig. 1.6. Therefore, the V2X communication for collision avoidance systems should be evaluated at full length for different driving parameters, vehicle-related properties as well as different traffic conditions before being deployed in real-life traffic environments.

1.4   COMBINED LATERAL AND LONGITUDINAL MOTION CONTROL FOR COLLISION AVOIDANCE

One of the most challenging missions for a collision avoidance system is to know how to properly and wisely use the available actuators of vehicle to prevent a collision. Most research done in collision avoidance employs ABS as the main actuation for controlling a vehicle's longitudinal dynamics. However, in emergency scenarios when an obstacle (such as a deer), suddenly appears in the middle of the road, it is difficult to avoid collision with the obstacle even though the vehicle has fully braked and made full use of the road friction [24]. Therefore, incorporating other actuators into the lateral control system may improve the performance of collision avoidance system.

Studies on collision avoidance have clearly indicated that braking is more efficient at lower speeds and steering is more efficient at higher speeds [25]. However, this does not mean that pure braking or pure steering always provides the most efficient form of collision avoidance. Due to complex traffic conditions and various road frictions, the emergency brake may cause a vehicle to lose its stability, but lateral motion control is helpful to solve this problem in the design of longitudinal collision avoidance control strategy. Therefore, combined steering and braking may be a more efficient way to avoid collision over a wide range of speed, especially when vehicle speed is high and the dynamics are nonlinear [26].

By either automatic steering or braking, accidents can be avoided or mitigated to a certain extent by using some active safety systems currently available on the market. However, there is still no collision avoidance system that combines lateral and longitudinal control technologies on mass-produced vehicles, this is because these functions are often tailored for specific accident types and for each type either braking or steering may be possible. In addition, the lateral and longitudinal motion control technologies used for collision avoidance still face the following challenges.

The first challenge that this technology faces is to decide on when to intervene with automatic braking and when to intervene with automatic steering, as shown in Fig. 1.7. Decisions on when and how to assist the driver in avoiding collisions are based on so-called threat assessment of traffic scenarios [27]. It mainly depends on the collision avoidance system's ability to perceive and understand the surrounding traffic conditions and the ability to control the vehicle, e.g., how fast the system can steer or brake automatically. Another challenge is to determine the last time before a possible conflicts situation, which can be obtained by analyzing the driving behavior of experienced drivers under critical traffic conditions. In cases where a collision cannot be avoided but only mitigated, the decision is much subtler since it is influenced by many aspects.

As a core component of the collision avoidance system, design of controllers at the vehicle level is another challenge in the effective implementation of collision avoidance objectives. From the existing literature, we found that most of the research on vehicle control has historically developed in two completely separate motion directions, with little consideration given to vehicle characters such as dynamic coupling and tire force coupling, as shown in Fig. 1.8. In these studies, longitudinal motion control has been primarily aimed at the design of spacing controllers that maintain a desired vehicle-to-vehicle spacing within a platoon of vehicles. In contrast, lateral motion control has been directed exclusively toward the problem of steering the vehicle into adjacent lanes to avoid collisions with obstacles. Nevertheless, it is known that the motions are not completely independent. Consequently, unidirectional control designs invite unforeseen dangers arising from its neglect of the coupling dynamics. Moreover, the control targets chosen in these studies are often mutually exclusive, with little thought given to possible

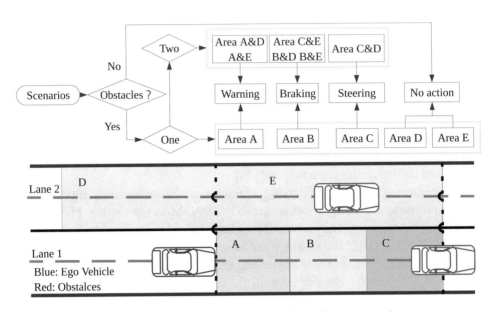

Figure 1.7: Combined lateral and longitudinal control for collision avoidance.

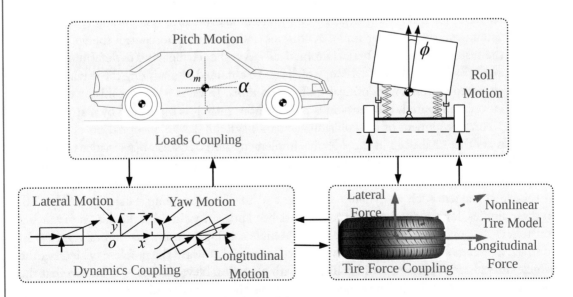

Figure 1.8: Dynamics coupling of vehicle model.

benefits derived from complementary coverage in both directions. Thus, in an effort to address some of these shortcomings of the existing approach, the collision avoidance controller should be designed to provide a unified approach to vehicle modeling, control, and reference in both longitudinal and lateral motion direction.

1.5   CONTRIBUTIONS OF THIS BOOK

Collision avoidance systems for intelligent connected vehicles require integration of path-planning and vehicle control subsystems. This book presents contributions in the areas of improving vehicle's active safety by summarizing and designing collision avoidance systems in lateral and longitudinal motion directions. An outline of the book and its contributions are presented below.

Chapter 2 provides a review of literature relevant to different approaches to solve the path-planning for collision avoidance. The global and local path-planning algorithms are introduced to create an optimal possible path from initial point to the destination in the complex traffic environment while satisfying the criteria set. For better understanding, several illustrating examples and some implementations in MATLAB scripts are given.

Chapter 3 provides a review on path-tracking algorithms in the control phase of the autonomous driving vehicles. In this chapter, traditional control method, human-like control strategies, and intelligent approach are discussed to solve the path-tracking problem and ensure the vehicle successfully avoids the collisions. The path-tracking controllers proposed in this chapter are simulated and verified by MATLAB software.

Chapter 4 poses the path-planning problem for collision avoidance considered in this section. An improved artificial potential field is established according to the vehicle kinematics and parameters of road, and then a local path for collision avoidance is obtained. Furthermore, a nonlinear bicycle model is presented, and the optimal planning problem is solved numerically for this model to demonstrate properties of the corresponding optimal trajectories.

Chapter 5 mainly discusses the path tracking for automated vehicle in complex conditions. We present a path-tracking control architecture, which includes the modeling of the path-tracking and vehicle dynamic, system uncertainty analysis and the design of the robust gain-scheduling lateral tracking controller. Several case studies are also proposed to verify the tracking performance of the controller.

Chapter 6 introduces the basic principles and methods of FCW and AEB systems as well as their specific applications in longitudinal active collision avoidance. In addition, the application of collision avoidance system based on V2X communication in the prevention of multi-vehicle pile-up is analyzed.

In Chapter 7, conclusions are presented along with suggestions for future work.

CHAPTER 2

Path-Planning Algorithms for Collision Avoidance

Path-planning is a widely studied subject in the field of unmanned robots, and it is also one of the basic operations required to realize collision avoidance for autonomous driving vehicles. With information about obstacles location given by onboard sensors and road geometry information provided by highly automated driving map [28], a path planner calculates a feasible collision-free trajectory to the destination in the complex traffic environment while satisfying the criteria set. The path planning problem for collision avoidance can be described using the following set of specifications.

The final result of path planning typically consists of a path or a trajectory, $\tau : [t_0, t_f] \to X$, such that $\tau(t_0) = x_0$ and $\tau(t_f) = x_f \in X_g$, where X is the vehicle's state space, and X_g is the goal region, x_0, x_f are the initial and final state in the path, for some $t_0 \in \mathbb{R}$ and $t_f \in \mathbb{R}$ defined by the planner such that $t_0 \leq t_f$. In the case where a planning algorithm associates a time with each $\tau(t)$ for all $t \in [t_0, t_f]$, τ is then referred to as a trajectory [29].

Generally, the path-planning methods can be grouped into global planner and local planner, global planner uses a priori information of the road map to create an optimal possible path to the destination, it is good in producing a global path that is either time optimal or distance optimal, but poor in reacting to unknown obstacles. In contrast, local planner recalculates a dynamic path to avoid collision with other vehicles or obstacles, it works well in dynamic and initially unknown environment, but is inefficient especially in a complex environment [30]. Next, we summarize the main approaches to solve these two kinds of path-planning problems.

2.1 COMMONLY USED GLOBAL PATH-PLANNING ALGORITHMS

As said above, the purpose of the global path-planning algorithm is to find a short and safe path from the current location to the destination, while meeting the global planning goals and adapting to the updated traffic conditions. For intelligent-connected vehicles, global path planner can obtain the traffic situation of the entire city through V2X communication and plan a safer path in advance, so that the vehicle can avoid being in dynamic, crowded, or noisy environment and effectively reduce the occurrence of vehicle collision accidents on road. The global path planner requires not only all available priori knowledge of the traffic information and road environment, but also some assumptions or forecasts on how the traffic conditions will evolve up over time, and these benefits come at the cost of higher computational complexity and more complete environment information.

In the design of the global path planner, the vehicle is generally regarded as a moving particle, the impact of vehicle's dynamic characteristics and structure size on collision avoidance performance is seldom considered. Therefore, global path planning is not the main topic of this book. Please refer to References [31–33] for more detailed introduction and application of this method. In this section, most commonly used search algorithms in global path planning are introduced, starting with simple Dijkstra algorithms and continue with advanced A* algorithms that can guarantee the optimal path and ending with RRTs algorithm that can include heuristics to narrow the search area or guarantee completeness. For better understanding, several illustrating examples and some implementations in MATLAB scripts are given.

(1) Dijkstra Algorithm

Because of the simplicity for debugging and its good performance on the experiments, the Dijkstra algorithm is considered to be one of the simplest and effective methods to solve global path planning. For a given source node in the graph, the algorithm finds the shortest path from the source node to a target node in a weighted graph, the graph can either be directed or undirected. One stipulation to using the algorithm is that the graph needs to have a non-negative weight on every edge [34, 35].

The first step of the Dijkstra algorithm is to represent the free space in the environment with a graph and use a graph search algorithm to compute a path. The cellular decomposition method partitions environmental free space into cells and creates a graph based on cell connectivity [36]. The road map method chooses a series of points in the environment and defines a graph based on lines or paths connecting these points.

For example, assuming that all continuous paths are feasible, try to find the shortest path from home to school by using the Dijkstra algorithm, as shown in Fig. 2.1. Dijkstra uses the road map approach to convert the road network into a visibility graph according to the information of a grid cell map, and the visibility graph can be used to find minimum distance piecewise linear paths using graph search algorithms.

This method starts with a set of candidate nodes where the vehicle is able to navigate (free space), and then assigns a cost value to each of them. From the starting point (home), this value is increased by the necessary number of nodes to pass through to reach each node. Take the case mentioned above for example, given a source node P_1 in the graph, find shortest paths from source to node P_8 in the given graph. At first, a shortest path tree is built with given nodes as root. We maintain two sets, one set, S_n, contains nodes included in shortest path tree, and other set, T_n , includes nodes not yet included in shortest path tree, as shown in Table 2.1. At every step of the algorithm, we find a node which is in the set T_n and has a minimum distance from the source. The minimum value of the sum of all nodes from the source node and the ending node

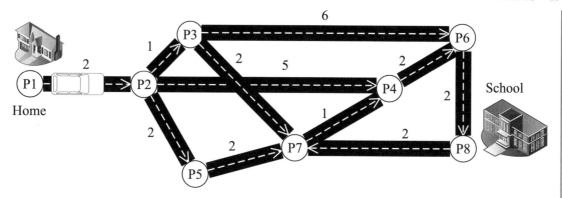

Figure 2.1: A weighted graph for the Dijkstra algorithm.

Table 2.1: The process of finding the shortest path from node P_1 to node P_8

Step	Set: S_n	Set: T_n	Nodes	Shortest Distance [1] [2] [3] [4] [5] [6] [7] [8]
1	P_1	$P_2, P_3, P_4, P_5, P_6, P_7, P_8$	P_2	0, 2, ∞, ∞, ∞, ∞, ∞, ∞
2	P_1, P_2	$P_3, P_4, P_5, P_6, P_7, P_8$	P_3	0, 0, 3, 7, 4, ∞, ∞, ∞
3	P_1, P_2, P_3	P_4, P_5, P_6, P_7, P_8	P_5	0, 0, 0, 7, 4, 9, 5, ∞
4	P_1, P_2, P_3, P_5	P_4, P_6, P_7, P_8	P_7	0, 0, 0, 7, 0, 9, 5, ∞
5	P_1, P_2, P_3, P_5, P_7	P_4, P_6, P_8	P_4	0, 0, 0, 6, 0, 9, 0, ∞
6	$P_1, P_2, P_3, P_4, P_5, P_7$	P_6, P_8	P_6	0, 0, 0, 0, 0, 8, 0, ∞
7	$P_1, P_2, P_3, P_4, P_5, P_7, P_8$	P_8	P_8	0, 0, 0, 0, 0, 0, 0, 10
8	$P_1, P_2, P_3, P_4, P_5, P_6, P_7, P_8$			0, 0, 0, 0, 0, 0, 0, 0

is the shortest path. After the success of finding the global path from the start to the goal, all the selected nodes are translated into positions in the reference axes as the form $P_i = (x_i, y_i)^T$.

An implementation of a Dijkstra algorithm in MATLAB is given in Listing 2.1. The result of the algorithm on the map is visualized in Fig. 2.2.

The shortest path between nodes 1 and 8, which could be found using Dijkstra algorithm, is $P = 1(\text{Home}) \rightarrow 2 \rightarrow 3 \rightarrow 7 \rightarrow 4 \rightarrow 6 \rightarrow 8(\text{School})$, this path has a total length of $d = 10$.

(2) A* Algorithm

A* (A star) is one of the best and popular algorithms for finding the shortest path between two locations in a mapped area. It is a graph traversal and informed search algorithm because it includes additional information or heuristic [37]. The heuristic function estimates the cost of

Listing 2.1: Dijkstra algorithm

```
% source nodes s
s = [1 2 2 2 3 3 4 5 5 6 7 8];
% source nodes t
t = [2 3 4 5 6 7 6 6 7 8 4 7];
% the distances along the edges
weights = [2 1 5 2 6 2 2 5 2 2 1 2];
% Create and plot a graph with weighted edges
G = digraph(s,t,weights);
% custom node coordinates
x = [0 1 2 2 2 3 3 4];
% custom node coordinates
y = [0 0 1 0 -1 1 -1 0];
% Plot a graph using custom node coordinates
p = plot(G,'XData',x,'YData',y,'EdgeLabel',G.Edges.Weight);
% Find the shortest path between nodes 1 and 8
[P,d] = shortestpath(G,1,8);
% Highlight this path in red
highlight(p,P,'EdgeColor','r');
```

the path from the current node to the goal node, it can be

$$
d = \begin{cases}
|x_i - x_j| + |y_i - y_j| & \textit{Manhattan distance} \\
\sqrt{(x_i - x_j)^2 + (y_i - y_j)^2} & \textit{Euclidean distance} \\
\max\left(|x_i - x_j|, |y_i - y_j|\right) & \textit{Diagonal distance.}
\end{cases}
\tag{2.1}
$$

In some simple cases, the Euclidean distance is often used to compute this cost for path first [38]. The cost is increased to make a priority in the forward direction to the goal in the part of the graph tree that has not been explored yet. This enables the A* algorithm to distinguish between more or less promising nodes, and consequentially it can accelerate the search speed when searching for the shortest path for motion in a wide area of outdoor environments.

Figure 2.3 shows a simple example to demonstrate the A* algorithm: suppose someone wants to move from point A to point B in the most efficient way possible, but a wall and an obstacle separates the two points. The first step of path planning is to simplify the research area, which is reduced to a simple two-dimensional array. Each item in the array represents one of the squares on the grid, and its status is recorded as open or closed. Next, A* algorithm is adopt to figure out which squares we should take, and then connect the nodes of each of the squares

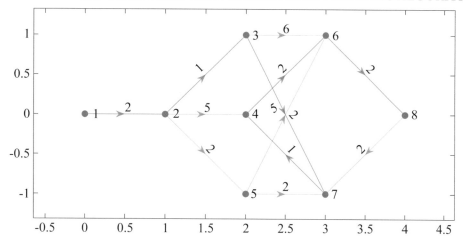

Figure 2.2: The shortest path from home to school.

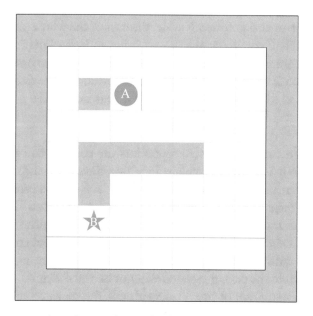

Figure 2.3: Two-dimensional grid map for path planning.

to form the shortest path from point A to point B. During the path search, an open nodes list and a closed nodes list are built: the open nodes list is a list of all locations immediately adjacent to areas that have already been explored and evaluated, but the closed nodes list is a record of all locations which have been explored and evaluated by the algorithm. Starting from a given node of a graph, the A* algorithm computes heuristic functions to minimize the cost estimate (least

Listing 2.2: The pseudocode of A* algorithm

---

(1) $g(s_{start})\leftarrow 0$ and g-values of the rest of the states are set to ∞

(2) $OPEN \leftarrow \{s_{start}\}$, CLOSED $\leftarrow \Phi$

1 ComputePath ( )

2 while (s_{goal} is not expanded)

3 removes with the smallest $f(s)$ from $OPEN$

4 for each successors of s

5   if $g(s') > g(s) + c(s, s')$

6   $g(s') \leftarrow g(s) + c(s, s')$

7   insert/update s' in   with $f(s') \leftarrow g(s') + h(s')$

---

distance travelled) for the path from this node to the given goal. It does this by maintaining a tree of paths originating at the start node and extending those paths one edge at a time until its termination criterion is satisfied [39]. Listing 2.2 shows the pseudocode of A* Algorithm.

During algorithm execution for each node, the cost of the whole path is calculated to consist of cost-to-here $g(n)$ and cost-to-goal $h(n)$, as shown in Equation (2.2). At each iteration of its main loop, A* algorithm needs to determine which nodes to extend based on the cost estimate required searching the path all the way to the goal. According to the defining characteristics of the A* algorithm mentioned above, selects the node n that minimizes

$$f(n) = g(n) + h(n), \tag{2.2}$$

where n is the next node on the path, $g(n)$ represents the cost (distance) of the path from the starting node to n, and it is easy to calculate in a two-dimensional grid, the values of $g(n)$ will increase as we get farther away from the starting node. $h(n)$ is a problem-specific heuristic function that estimates the cost of the shortest path from n to the goal. The closer the estimated movement cost is to the actual cost, the more accurate the final path will be. If the heuristic function is admissible, meaning that it never overestimates the actual cost of reaching the goal and ensures that a path with the least cost is returned from the beginning to the goal; but if the estimate is off, it is possible the path generated will not be the shortest.

The typical implementation of the A* algorithm is the repeated selection of nodes to expand with a minimum cost, as shown in Fig. 2.4. The algorithm operation is as follows. In the beginning, the open list O contains only the starting node n_1, which has zero cost-to-here $g(n)$ and is without a connection to a previous node. Next, the starting node n_1 is removed from the open list and added to the closed nodes list. At the same time, put all drivable nodes $n_{(x,y)}$ adjacent to current position into the open list. If the heuristic $h(n)$ satisfies the additional condition $h(x) \leq d(x, y) + h(y)$ for every drivable nodes $n_{(x,y)}$, then $h(n)$ is called monotone, or consistent. With a consistent heuristic, A* algorithm is guaranteed to find an optimal path without processing any node more than once. For each node $n_{(x,y)}$ in drivable adjacent tiles,

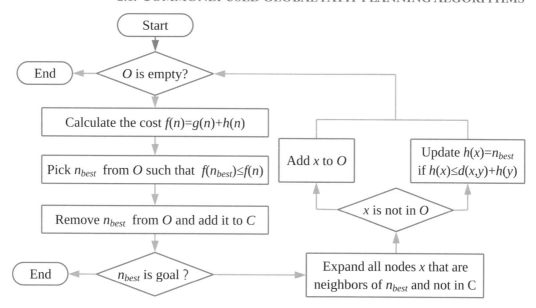

Figure 2.4: The flow diagram of A* algorithm.

ignore this node if it is in a closed list. If $n_{(x,y)}$ is already in the open list, then check if the $f(n)$ score is lower when we use the current generated path to get there. If it is, the node with the lowest $f(n)$ value is removed from the open list, the $f(n)$ and $g(n)$ values of its neighbors are updated accordingly, and these neighbors are added to the open list. The algorithm continues until a goal node has a lower $f(n)$ value than any node in the queue. The $f(n)$ value of the goal is then the cost of the shortest path, since $h(x)$ at the goal is zero in an admissible heuristic.

In this example, shown in Fig. 2.3, we assign a cost of one to each horizontal or vertical node moved, and the values for $f(n) = g(n) + h(n)$ are listed according to the following:

- $f(n)$ (score for square): Top left corner

- $g(n)$ (cost from A to square): Bottom left corner

- $h(n)$ (estimated cost from square to B): Bottom right corner

The Manhattan distance in Equation (2.1) is adopted to calculate the values of $g(n)$ and $h(n)$. According to the definition of each parameter in Equation (2.2). The value of $g(n)$ can be obtained by counting the nodes on the determined path from the starting node to n, and the value of $h(n)$ can be achieved by calculating the total number of nodes moving horizontally and vertically from the current node to the target node, ignoring the diagonal movement and any obstacles that may exist in the path. The proposed algorithm only gives the length of the shortest path. In order to find the actual global path, the algorithm can be modified so that each

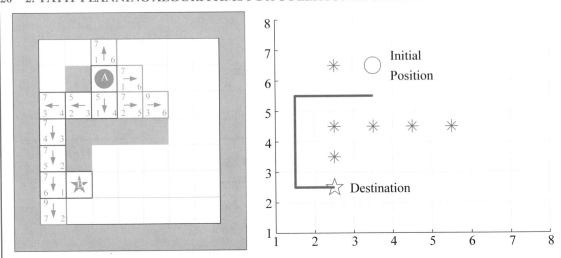

Figure 2.5: Simulation results of A* algorithm.

node on the path can track its parent nodes. After this algorithm is run, it can make the end node point to its parent node until the predecessor of a node becomes the start node, as shown in Fig. 2.5.

(3) RRT and RRT* Algorithms

Sampling-based planning algorithms such as RRT and RRT* are widely used in path-planning for autonomous vehicles in recent decades [40]. They are probabilistic complete algorithms and have natural support for searching nonconvex and high-dimensional spaces by randomly building a space-filling tree. The RRT and RRT* algorithm plant a tree rooted at the starting node, then the tree is constructed incrementally by randomly selecting samples in the search space and is inherently inclined to grow towards the unexplored area. Generally, RRTs can be considered a Monte Carlo method for generating open-loop trajectories for autonomous vehicles with state constraints, and it can be incorporated into the development of a variety of different planning algorithms [41].

RRT is a common option that both creates a graph and finds a path, and the path will not necessarily be optimal. But RRT* is an optimized modified algorithm that aims to achieve a shortest path, whether by distance or other metrics, it is an extension of RRT with faster convergence as compared to its predecessors [42]. Both algorithms are good at handling problems with nonholonomic or kinematic constraints, and they are implemented in MATLAB and observed in this section. To demonstrate the idea, the algorithms will be implemented in a 2D space with bounds. However, both algorithms can be built into any real continuous dimensional space.

The key idea of RRT algorithm is to bias the exploration toward unexplored portions of the space by sampling numerous points randomly. An RRT grows a tree of dynamically feasible

Listing 2.3: The pseudocode of RRT algorithm

```
function BuildRRT(q_init, K, Δq)
   T.init(q_init)
   for k = 1 to K
      q_rand = Sample()  -- chooses a random configuration
      q_nearest = Nearest(T, q_rand) -- selects the node in the RRT tree that is closest to qrand
      if  Distance(q_nearest, q_goal) < Threshold then
         return true
      q_new = Extend(q_nearest, q_rand, Δq)  -- moving from q_nearest an incremental distance in the
direction of q_rand
      if q_new ≠ NULL then
         T.AddNode(q_new)
return false
function Sample() -- Alternatively,one could replace Sample with SampleFree(by using a colli-
sion detection algorithm to reject samples in C_obstacle
   p = Random(0, 1.0)
   if 0 < p < P_rob then
      return q_goal
   elseif P_rob < p < 1.0 then
      return RandomNode()
```

trajectories and incrementally pulls the search tree toward unexplored area. After a sample is drawn, it attempts to connect to the nearest node in the tree. If the established connection between two nodes can pass through the free space freely and does not exceed any constraints, the sample is added to the tree and will become a node of the tree and the line between this new node and the nearest node of the tree will become an edge and also added to the tree. These steps are repeated numerous times until the endpoint is added to the tree. By tracing all the nodes, a continuous path is constructed from the starting node to the end point, thereby successfully avoiding all obstacles. Listing 2.3 shows the pseudocode of RRT algorithm.

Most practical implementations of RRTs lead the tree to the goal by increasing the probability of the sampling nodes from a specific area and limiting the incremental length of the connection between the tree and a new node. In the case of uniform sampling in the search space, the extension probability of the nodes in the tree is proportional to the size of the Voronoi region. The higher the probability, the more obvious the tree's direction to grow toward the goal. Otherwise, the length of the connection often needs to be adjusted appropriately according to the random samples. If the distance between the random sample and the nearest node in the tree exceeds its limitation, a new node at the maximum distance from the tree along the line to the random sample is used instead of the random sample itself.

The benefit of RRT algorithm is its speed and implementation, but the basic RRT algorithm also has some shortcomings, the most obvious of which is that if there are a lot of obstacles in the space, the tree is difficult to pass through a narrow channel, and it is difficult to find and connect the target point between two obstacles. Most of these disadvantages can be fixed with additional extras to the basic RRT algorithm. For example, RRT* is also a sampling-based planning method, but is an optimized version of RRT. Both RRT and RRT* converges to an optimal solution if sufficient running time is provided. But when the distribution of obstacles in space is very complex, the number of nodes obtained by RRT algorithm tends to infinity, while RRT* algorithm will provide the possible shortest path to the target. The rationale of RRT* is the same as RRT, but two additional steps are added to the traditional RRT algorithm to make it more widely used. First, RRT* examines nearby vertices to first find the optimal parent to the newly added vertex. After finding the nearest node in the graph, check the vertices on the neighborhood of the new node with a fixed radius. If it is found that the cost of a node is less than that of the proximal node, then the proximal node is replaced with the lower-cost node. Second, RRT* rewires edges of neighboring vertices through the newly added vertex. When a vertex connects to the cheapest neighbor, it examines to see if the neighbor rewires to the newly added vertex to reduce their cost. If the cost is indeed decrease, the neighbor is reconnected to the newly added vertex. Listing 2.4 shows the pseudocode of RRT* algorithm. In fact, the two mature path-planning algorithms can be directly called by commands (plannerRRT and plannerRRTStar) in MATLAB at present.

As described earlier, there are many algorithms that can be applied to solve the simple problem of how to get from the starting point to the goal in a 2D cluttered environment. Next, RRT and RRT* algorithms will be respectively used to solve a specific problem, and MATLAB programming will be used to demonstrate the simulation results of these two algorithms.

The test cases consist of predefined environments with a specific layout, designed to test different aspects of RRT and RRT* algorithms. The task for the vehicle is to travel from a start point to a given goal in the most efficient way. The simulated vehicle is not provided with any information about the environment at the start, so it will therefore need to explore the map and react to obstacles which are discovered on the way to the goal. The map for simulation, see Fig. 2.6, consists of a series of obstacles which are slabs erected on the floor. The path is planned by building a tree starting from the position of the vehicle. RRT and RRT* methods are then adopted to increase the number of nodes in the tree to effectively search the non-convex space and guide the vehicle to the goal continuously. Each time after a node is added to the tree, it is checked if the goal can be reach in from the added node. If the goal position is reachable, the goal position is added to the tree with the recently added node as its parent. At this point, the path planning is complete. An example of a completed path is shown in Fig. 2.6.

Listing 2.4: The pseudocode of RRT* algorithm

```
Rad = r
G(V,E) //Graph containing edges and vertices
For i_tr in range(0…n)
    X_new = RandomPosition()
    If Obstacle(X_new) == True, try again
    X_nearest = Nearest(G(V,E),X_new)
    Cost(X_new) = Distance(X_new,X_nearest)
    X_best,X_neighbors = findNeighbors(G(V,E),X_new,Rad)
    Link = Chain(X_new,X_best)
    For x' in X_neighbors
        If Cost(X_new) + Distance(X_new,x') < Cost(x')
            Cost(x') = Cost(X_new)+Distance(X_new,x')
            Parent(x') = X_new
            G += {X_new,x'}
    G += Link
Return G
```

As can be seen from Fig. 2.6, the graphs obtained by using RRT* are characteristically different from those of RRT. The RRT* algorithm exhibits paths which have been rewired so that they are straighter and for this case then closer to optimal. For finding an optimal path, especially in a dense field of obstacles, the RRT* algorithm has been shown to find a path which will in the limit of infinitely many nodes almost surely converge to the optimal solution. An implementation of RRT and RRT* algorithms in MATLAB is given in Listing 2.5.

2.2   TRAJECTORIES GENERATION FOR A LOCAL PATH PLANNER

Global path-planning generates a relatively safe route from the starting point to the destination before the vehicle starts its motion, so that the vehicle can avoid being in a crowded or dangerous traffic environment. In order to transform the global path into a local trajectory suitable for vehicle movement, the local planner needs to consider the geometric parameters of the road, the movement of other vehicles and the constraints of the autonomous vehicles, finally recalculates a local trajectory that can avoid collision in the complex traffic condition [43].

For local path planner, it is not possible to use the whole map because the sensors are unable to update the map in all regions and a large number of nodes would raise the computational burden. Therefore, in order to recalculate the trajectory in a specific road environment to

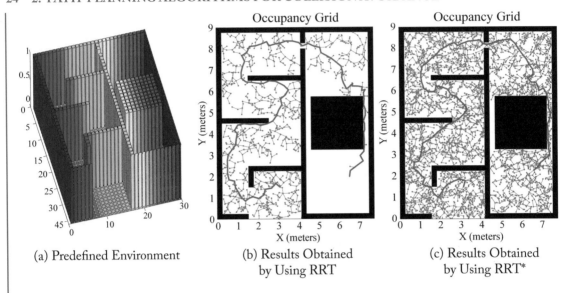

(a) Predefined Environment   (b) Results Obtained   (c) Results Obtained
by Using RRT   by Using RRT*

Figure 2.6: Simulation results of RRT and RRT* algorithm.

avoid random obstacles, a local map surrounding of the vehicle is built based on the information from on-board sensors and is updated as the vehicle moves. Then, with the updated local map and the provided global path, the local path planner generates a virtual desired trajectory to the destination and tries to match the trajectory as much as possible to the provided path from the global planner [44]. In addition, both passenger's ride comfort and transition time should be considered in the design of virtual desired trajectory for collision avoidance. Different local path-planning approaches in consideration of transition time, allowable lateral acceleration and vehicle speed have been proposed in this chapter, the shapes of these trajectories could be arcs and segments, Clothoids lines [45], B-spline curves [46], Bezier curves [47], or optimal curves [48]. All of them create intermediate waypoints following the generated trajectory.

(1) Circular Trajectory
The circular trajectory is composed of two arcs and a straight-line segment, wherein two arc curves are connected with the center line of the initial lane and its adjacent lane respectively, but the straight-line segment connects the two arcs, as shown in Fig. 2.7.

These parameters of arcs and line must satisfy the geometric constraint of road as follows.

$$L_w = 2\frac{V^2}{a_{\max}}\left[1 - \cos\left(\frac{a_{\max}t_c}{V}\right)\right] + Vt_s \sin\left(\frac{a_{\max}t_c}{V}\right), \tag{2.3}$$

Listing 2.5: RRT and RRT* algorithms

```
ss = stateSpaceSE2;
sv = validatorOccupancyMap(ss);
load EnvironMap.mat
map = occupancyMap(EnvironMap,4);
sv.Map = map;
sv.ValidationDistance = 0.05;
ss.StateBounds = [map.XWorldLimits; map.YWorldLimits; [-pi pi]];
planner01 = plannerRRT(ss,sv);
planner01.MaxConnectionDistance = 0.3;
planner02 = plannerRRTStar(ss,sv);
planner02.ContinueAfterGoalReached = true;
planner02.MaxIterations = 5000;
planner02.MaxConnectionDistance = 0.3;
start = [3,1.5,0];
goal = [6,2,0];
rng(300,'twister'); % for repeatable result
[pthObj01,solnInfo01] = planner01.plan(start,goal);
[pthObj02, solnInfo02] = plan(planner02,start,goal);
figure(1)
map.show;hold on;
plot(solnInfo01.TreeData(:,1),solnInfo01.TreeData(:,2),'.-');
plot(pthObj01.States(:,1), pthObj01.States(:,2),'r-','LineWidth',2)
figure(2)
map.show;hold on;
plot(solnInfo02.TreeData(:,1),solnInfo02.TreeData(:,2), '.-');
plot(pthObj02.States(:,1),pthObj02.States(:,2),'r-','LineWidth',2);
```

where L_w is the width of a lane, V is the velocity of vehicle in longitudinal direction, and $a_{\max}$ is the maximum allowable acceleration in lateral direction. t_c and t_s are the time that corresponds to the circular portion and the straight-line portion of the trajectory.

It can be seen from Fig. 2.7 that the total transition time of collision avoidance maneuver is

$$T = t_s + 2t_c. \tag{2.4}$$

Supposed that L_w, V, and $a_{\max}$ have been determined in certain scenarios, according to Equation (2.3), the total transition time T will decrease as the value of t_s decreases, as shown in

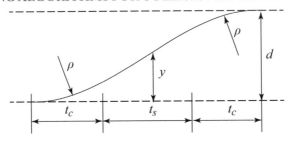

Figure 2.7: Definition of the variables for circular trajectory.

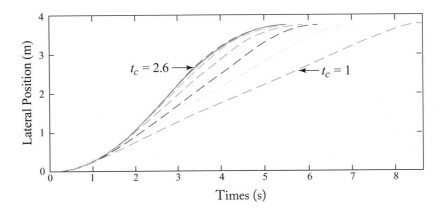

Figure 2.8: Planned trajectories with various circular portions.

Fig. 2.7. This means that the smaller the proportion of straight-line segment in the trajectory, the less transition time is required for collision avoidance. Although circular trajectory satisfies the acceleration constraint for ride comfort, but the jerk is infinity at the beginning and end of the trajectory, Fig. 2.8 shows the planning trajectories with various circular portions.

The MATLAB program for drawing the road trajectories under different curvature parameters is given in Listing 2.6.

(2) Sinusoid Approximation
Inspired by the shape of the sinusoid, an approximate trajectory for collision avoidance can also be constructed by adjusting the parameters of the sinusoidal curve. The equation of the trajectory can be described as follows,

$$y(t) = \frac{L_w}{2}\left[1 + \sin\left(\alpha t - \frac{\pi}{2}\right)\right], \quad 0 \le \alpha t \le \pi, \tag{2.5}$$

Listing 2.6: MATLAB program for circular trajectory

```
V=31.3;g=9.8;amax=0.05*g;d=3.75;
p=(V^2)/amax;
Tcir=2*sqrt(d/amax);
syms t;
ycir=p*(1-cos(V/p*t))*(t>=0&t<=(Tcir/2))+p*(1+cos(V/p*(Tcir-t))-2*cos(V/p*(T-
cir/2)))*(t>=(Tcir/2)&t<=Tcir);
fplot(ycir,[0 Tcir]);
xlabel('time(s)');ylabel('lateral position(m)');
hold on;
for i=1:0.2:Tcir/2
  Tcir1=2*i+(d-2*p)/(V*sin(V/p*i))+2*p/V*cot(V/p*i);
   ts=Tcir1-2*i;
ycir=p*(1-cos(V/p*t))*(t>=0&t<=i)+(p*(1-cos(V/p*i))+V*(t-i)*tan(V/
p*i))*(t>=i&t<=(i+ts))+...
   (p*(1+cos(V/p*(Tcir1-t))-2*cos(V/p*i))+V*ts*tan(V/p*i))*(t>(i+ts)&t<=Tcir1);
fplot(ycir,[0 Tcir1],'--');
hold on;
end
```

where α is a tuning parameter that keep the maximum lateral acceleration of vehicle on this trajectory satisfies the lateral acceleration limit, all other parameters are taken as the same values as that in previous section.

To satisfy the acceleration constraint, α is selected as $\sqrt{\frac{2a_{max}}{L_w}}$. The lateral acceleration and transition time of this trajectory is modified as

$$\ddot{y}(t) = -a_{\max} \sin\left(\alpha t - \frac{\pi}{2}\right), \quad 0 \le \alpha t \le \pi \qquad (2.6)$$

$$T = \pi \sqrt{\frac{L_w}{2a_{\max}}}. \qquad (2.7)$$

Based on the parameter values of vehicle and road in the previous section, a sinusoidal trajectory is established by using Equation (2.5), as shown in Fig. 2.9a. Comparing the sinusoidal trajectory and the circular trajectory, although the transition time of the sinusoidal trajectory is longer than that of the circular trajectory, the latter eliminates the discontinuity of the second derivative at the midpoint of the trajectory, as shown in Fig. 2.9b. The MATLAB program for sinusoid approximation simulation is given in Listing 2.7.

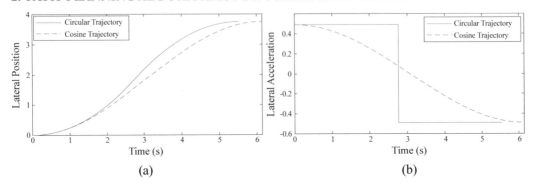

Figure 2.9: Comparison between a sine trajectory and a circular trajectory.

The two trajectories presented above have infinite jerk at a particular position, which is not an ideal and feasible path for collision avoidance from the perspective of tracking performance and controller design.

(3) Polynomial Trajectory
In this section, the expected trajectory from initial point to destination point is defined as a set of position coordinates as functions of time, and velocity and acceleration along the trajectory can be computed by differentiating position with respect to time. A common way to build such a trajectory is to deal with polynomial fitting. Given a set of necessary points and vehicle constraints, a smooth trajectory can be specified using cubic or five order polynomials which have independent coefficients [49].

The path between two points is defined with a cubic polynomial,

$$Y(t) = a_0 + a_1 t + a_2 t^2 + a_3 t^3. \tag{2.8}$$

Velocity and acceleration can be calculated by differentiating the above equation:

$$\dot{Y}(t) = v(t) = a_1 + 2a_2 t + 3a_3 t^2 \tag{2.9}$$

$$\ddot{Y}(t) = \alpha(t) = 2a_2 + 6a_3 t. \tag{2.10}$$

The coefficients of the polynomials (i.e., a_0, a_1, a_2, a_3) are determined by satisfying the road constraints and the vehicle's constraints at initial time t_0 and end time t_f. The above coefficients can be solved by the following matrix equation:

$$\begin{bmatrix} 1 & t_0 & t_0^2 & t_0^3 \\ 0 & 1 & 2t_0 & 3t_0^2 \\ 1 & t_f & t_f^2 & t_f^3 \\ 0 & 1 & 2t_f & 3t_f^2 \end{bmatrix} \begin{bmatrix} a_0 \\ a_1 \\ a_2 \\ a_3 \end{bmatrix} = \begin{bmatrix} Y_0 \\ V_0 \\ Y_f \\ V_f \end{bmatrix}, \tag{2.11}$$

Listing 2.7: MATLAB program for sinusoidal approximation simulation

```
V=31.3;g=9.8;amax=0.05*g;d=3.75;
p=(V^2)/amax;
Tcir=2*sqrt(d/amax);
syms t;
ycir=p*(1-cos(V/p*t))*(t>=0&t<=(Tcir/2))+p*(1+cos(V/p*(Tcir-t))-2*cos(V/p*(T-
cir/2)))*(t>=(Tcir/2)&t<=Tcir);
fplot(ycir,[0 Tcir]);
hold on;
a=sqrt(2*amax/d);
Tcos=pi*sqrt(d/(2*amax));
ycos=d/2*(1-cos(a*t));
fplot(ycos,[0,Tcos],'--');
legend('circular trajectory','cosine trajectory');
xlabel('time(s)');ylabel('lateral position');
hold off;
acir=(diff(diff(p*(1-cos(V/p*t)))))*(t>=0&t<=(Tcir/2))+(diff(diff(p*(1+cos(V/p*(T-
cir-t))-2*cos(V/p*(Tcir/2))))))*(t>=(Tcir/2)&t<=Tcir);
figure;
fplot(acir,[0 Tcir]);
hold on
vcos=diff(ycos);
acos=diff(vcos);
fplot(acos,[0,Tcos],'--');
legend('circular trajectory','cosine trajectory');
xlabel('time(s)');ylabel('lateral acceleration');
```

where the initial and end positions are Y_0 and Y_f, the initial and end velocities are V_0 and V_f, respectively.

The accelerations of trajectory will be discontinuous using the cubic polynomial, but this discontinuity of acceleration makes the jerk infinite at each via point, and causes an impulsive jerk in the motion of the vehicle. Therefore, a fifth-order polynomial is proposed to describe the trajectory between two points:

$$Y(t) = a_0 + a_1 t + a_2 t^2 + a_3 t^3 + a_4 t^4 + a_5 t^5 \qquad (2.12)$$

$$\dot{Y}(t) = a_1 + 2a_2 t + 3a_3 t^2 + 4a_4 t^3 + 5a_5 t^4 \qquad (2.13)$$

$$\ddot{Y}(t) = 2a_2 + 6a_3 t + 12a_4 t^2 + 20a_5 t^3. \qquad (2.14)$$

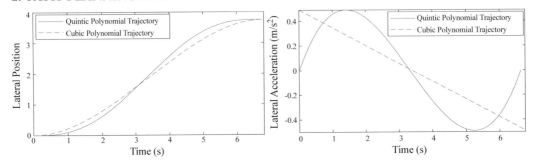

Figure 2.10: Polynomial trajectories and their lateral acceleration.

Three constraints (i.e., position, velocity, and acceleration) at each point can be described in a matrix form:

$$
\begin{bmatrix}
1 & t_0 & t_0^2 & t_0^3 & t_0^4 & t_0^5 \\
0 & 1 & 2t_0 & 3t_0^2 & 4t_0^3 & 5t_0^4 \\
0 & 0 & 2 & 6t_0 & 12t_0^2 & 20t_0^3 \\
1 & t_f & t_f^2 & t_f^3 & t_f^4 & t_f^5 \\
0 & 1 & 2t_f & 3t_f^2 & 4t_f^3 & 5t_f^4 \\
0 & 0 & 2 & 6t_f & 12t_f^2 & 20t_f^3
\end{bmatrix}
\begin{bmatrix}
a_0 \\ a_1 \\ a_2 \\ a_3 \\ a_4 \\ a_5
\end{bmatrix}
=
\begin{bmatrix}
Y_0 \\ V_0 \\ A_0 \\ Y_f \\ V_f \\ A_f
\end{bmatrix}.
\tag{2.15}
$$

The MATLAB program for designing and comparing polynomial trajectory is given in Listing 2.8.

The trajectories and their lateral acceleration are shown in Fig. 2.10.

(4) Trapezoidal Acceleration Trajectory

Although the higher-order polynomial is a simple and flexible method for path-planning, it is still hard to calculate a desired trajectory for collision avoidance in terms of lateral jerk. In order to quantitatively analyze the influence of lateral jerk on the path-planning the trapezoidal acceleration profile resultant trajectory is proposed and designed for collision avoidance. The key index of desired trajectory is to maintain passenger's ride comfort and limit the required transition time. For trapezoidal acceleration trajectory, the lateral dynamics of vehicle and the time requirement for collision avoidance are combined to meet the restriction of the transition time and ride comfort [50].

The lateral acceleration profile of the desired trajectory is plotted in a trapezoidal shape, and the slope of the acceleration profile is the lateral jerk of trajectory. A trapezoidal acceleration profile as shown in Fig. 2.11.

Listing 2.8: The MATLAB program for designing polynomial trajectory

```
V=31.3;g=9.8;amax=0.05*g;d=3.75;
m=(3-sqrt(3))/6;
xe=V*sqrt(d/amax*(60*m-180*m^2+120*m^3));
Tpol=xe/V;
syms t;
n=V*t/xe;
ypol=d*(10*n^3-15*n^4+6*n^5);
figure;
fplot(ypol,[0 Tpol]);
hold on;
xe1=V*sqrt(6*d/amax);
x=V*t/xe1;
ypol1=d*(3*(x^2)-2*(x^3));
Tpol1=xe1/V;
fplot(ypol1,[0 Tpol1],'--');
xlabel('time(s)');ylabel('lateral position(m)');
legend('quintic polynomial trajectory','cubic polynomial trajectory');
vpol=diff(ypol);
vpol1=diff(ypol1);
apol=diff(vpol);
apol1=diff(vpol1);
figure;
fplot(apol,[0 Tpol]);
hold on;
fplot(apol1,[0 Tpol1],'--');
xlabel('time(s)');ylabel('latral acceleration(m/s2)');
legend('quintic polynomial trajectory','cubic polynomial trajectory');
```

The lateral acceleration of trajectory $\ddot{y}_d$ can be written as

$$
\begin{aligned}
\ddot{y}_d = {} & J_{\max}t \cdot u\,(t) - J_{\max}\,(t - t_1) \cdot u\,(t - t_1) - J_{\max}\,(t - t_2) \cdot u\,(t - t_2) \\
& + J_{\max}\,(t - t_3) \cdot u\,(t - t_3) + J_{\max}\,(t - t_4) \cdot u\,(t - t_4) - J_{\max}\,(t - T) \cdot u\,(t - T),
\end{aligned}
\tag{2.16}
$$

where $J_{\max}$ denotes the maximum of jerk in lateral direction, $u\,(t)$ is the unit step function, t_1, t_2, t_3, t_4 are the temporal parameters for this trajectory, and T denotes the transition time, which is the elapse time of the collision avoidance maneuver. The lateral position of trajectory

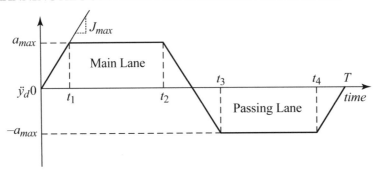

Figure 2.11: Trapezoidal acceleration profile.

can be obtained by integrating Equation (2.16) twice with respect to time:

$$y_d = \frac{J_{\max}}{6} \left\{ \begin{array}{c} t^3 \times u(t) - (t - t_1)^3 \times u(t - t_1) - (t - t_2)^3 \times u(t - t_2) \\ + (t - t_3)^3 \times u(t - t_3) + (t - t_4)^3 \times u(t - t_4) - (t - T)^3 \times u(t - T) \end{array} \right\}.$$
(2.17)

The temporal parameters for the collision avoidance trajectory satisfy the following constraints:

$$t_1 = \frac{a_{\max}}{J_{\max}}, \quad t_2 = \frac{-t_1^2 + \sqrt{t_1^4 + 4t_1 \frac{y_d}{J_{\max}}}}{2t_1},$$
(2.18)

$$t_3 = 2t_1 + t_2, t_4 = t_1 + 2t_2, \quad T = 2t_1 + 2t_2.$$

The transition time T is regarded as a function of lateral jerk and acceleration, t_2 and T can be written as,

$$t_2 = -\frac{1}{2}\frac{a_{\max}}{J_{\max}} + \frac{1}{2}\sqrt{\left(\frac{a_{\max}}{J_{\max}}\right)^2 + 4\frac{y_d}{a_{\max}}}$$
(2.19)

$$T = 2(t_1 + t_2) = \frac{a_{\max}}{J_{\max}} + \sqrt{\left(\frac{a_{\max}}{J_{\max}}\right)^2 + 4\frac{y_d}{a_{\max}}}.$$
(2.20)

Equations (2.17)–(2.20) are reintegrated to obtain the relationship between the lateral position, lateral acceleration, lateral jerk and transition time of the vehicle during active collision avoidance, as shown below.

$$L_{road} = \frac{J_{lc} t_1 t_2 T}{2} = \frac{T^2 \ddot{Y}_{\max}}{4} - \frac{\ddot{Y}_{\max}^2 T}{2 * J_{lc}}.$$
(2.21)

With same lateral offset distance, the MATLAB program in Listing 2.9 can be used to obtain the three-dimensional relationship surface among transition time, lateral acceleration limits, and lateral jerk limits, as shown in Fig. 2.12.

Listing 2.9: MATLAB program for calculating transition time

```
Llane=3.75;
Amin=0.1509;  Amax=3;  Ain=0.05;
Acc=Amin:Ain:Amax;
Jmin=0.6316; Jmax=5; Jin=0.05;
Jerk=Jmin:Jin:Jmax;
[X,Y]=meshgrid(Acc,Jerk);
TA=length(Acc);
TJ=length(Jerk);
Tlc=zeros(TA,TJ);
for I=1:TA
    for J=1:TJ
      Tlc(I,J)=Acc(I)/Jerk(J)+sqrt(Acc(I)^2/Jerk(J)^2+4*Llane/Acc(I));   % Calculation of
Transition Time
      if Tlc(I,J)<=2
         Tlc(I,J)=2;
      else
          Tlc(I,J)=Tlc(I,J);
      end
   end
end
mesh(X,Y,Tlc');
xlabel('Acceleration/(m/s^2)');
ylabel('Jerk/(m/s^3)');
zlabel('Transition time/(s)');
```

It can be seen from Fig. 2.11, that the transition time for collision avoidance gets shorter as the jerk limit and the lateral acceleration become larger, and the trapezoidal acceleration trajectory is reduced to the circular trajectory when the jerk is infinity.

Once the road width is known, the lateral offset distance can be determined. In addition, the maximum lateral acceleration and jerk can also be determined according to the ride comfort index of the vehicle, then one can obtain the virtual desired trajectory for collision avoidance from Equation (2.17).

Suppose a vehicle is driving at 20 m/s on a road with a lane width of 3.75 m. According to the different requirements of active collision avoidance, the simulation results of lateral position and lateral velocity of the expected collision avoidance path are obtained.

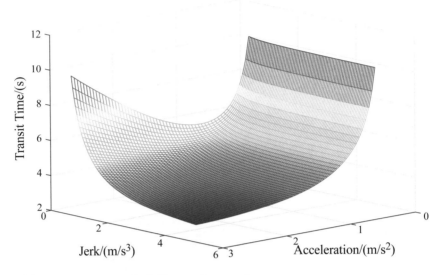

Figure 2.12: Transition time with different $J_{\max}$ and $a_{\max}$.

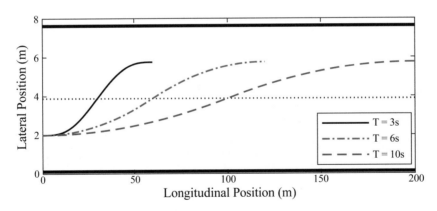

Figure 2.13: Trapezoidal trajectories with different T.

With same limits of jerk $J_{\max} = 5$ m/s^3 and different transition time $T = 3$ s, 6 s, 10 s, respectively, several trajectories are shown in Fig. 2.13.

With same transition time $T = 6$ s and different limits of lateral acceleration $a_{\max} = 0.5$ m/s^2, 1 m/s^2, 2 m/s^2, respectively, the lateral velocities of vehicle are shown in Fig. 2.14.

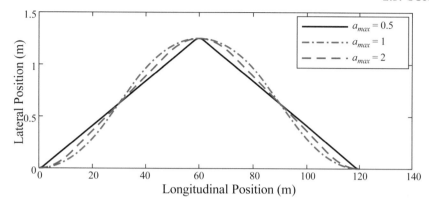

Figure 2.14: Trapezoidal trajectories with different $a_{\max}$.

2.3   SUMMARY

The development and application of collision avoidance system for intelligent vehicles requires a variety of the-state-of-the-art technologies, among which, path-planning algorithms play a critical role in improving safety and comfort for autonomous driving.

In this chapter, we summarized and compared substantial research on path-planning for collision avoidance system. Three types of global path planning algorithms, i.e., Dijkstra, A*, RRTs algorithm, are introduced to create an optimal possible path from initial point to the destination using a priori information of the road map. Then, different local path planners calculate a feasible collision-free trajectory to a destination in the complex traffic environment while satisfying the criteria set. Both kinds of path-planning methods allow the vehicle to meet the global planning objectives so as to adapt to updated traffic condition as well as the local safety objectives by recalculating a dynamic path to avoid collision.

Due to the limited performance of on-board sensors and computational units, most of the previous approaches do not consider the tracking control problem when solving the path-planning problem. More specifically, the high-level path planner often takes advantage of planning algorithms for robots based upon a simplified point-mass to compute a collision-free path in an arbitrary direction. Then the low-level path-tracking controller focuses on leading the vehicle onto the planned path using kinematic and/or dynamic lateral controllers. Since the vehicle can only move at the limits of its stability and handling capability in a constrained environment. Therefore, to solve the collision avoidance problems on the road, it is also necessary to design a reliable tracking controller considering the nonlinear dynamic characteristics of host vehicle and other moving vehicles that have these own motion properties. This issue will be discussed in the next chapter.

CHAPTER 3

Path-Tracking Algorithms for Collision Avoidance

Once the path for collision avoidance is predetermined by using the algorithms in the previous chapter, the autonomous vehicle is required to follow this specific trajectory given by an on-board planner and ensure that the vehicle successfully avoids the obstacles. This task can be accomplished by a path-tracking controller, which calculates the appropriate actuating input to control the vehicle's movement in both lateral and longitudinal directions. Specifically, its operations include adjusting the vehicle's lateral position by modifying the steering inputs, or adjusting the vehicle's longitudinal motion by modifying the brake or throttle actuators [51, 52]. A good path-tracking strategy timely tracks reference planned path by producing the required low-level control actions, and two critical effects for ground vehicles, dynamics, and kinematics should be accurately considered as an important factor in high-performance control systems, as shown in Fig. 3.1.

In a recent series of papers and the state of art, many control methods are emerged for solving path-tracking problem. In this chapter, a review on classical control strategies used for path-tracking controller in terms of the basic vehicle model is presented. This review will be focusing on trajectory tracking control in the control phase of the autonomous vehicles and the criteria methods used to evaluate the controller's performance in collision avoidance for evaluation and comparison purposes. The path-tracking controllers proposed in this chapter are simulated and verified by MATLAB software.

3.1   PID CONTROL

The PID control method is a typical representative of classical control algorithms for path-tracking, which has several advantages such as straightforward structure, good control performance, robust design, and simple implementation. The tracking performance of controller should be fast enough in real time without much overhead that increases computational burden and may suffer from stability problems, but off-line parameters tuning of PID controller is superior to the adaptive or intelligent techniques in this regard [53]. Therefore, PID controller was widely used in the control design and experimental verification of path-tracking problems.

PID controller consists of three terms and usually triggered by the error between actual response and the desired inputs. The three terms are Proportional (P), Integral (I), and Derivative

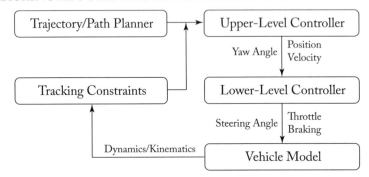

Figure 3.1: Scheme of the path-tracking controller.

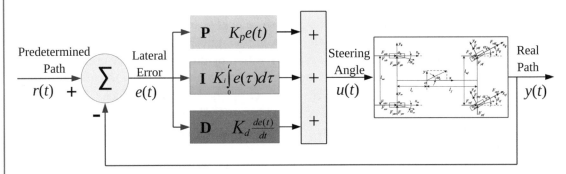

Figure 3.2: System structure of a PID controller.

(D), which corresponds to the action that each term applied to the triggering error, respectively. The output of PID controller can be expressed by Equation (3.1):

$$u\left(t\right) = K_P e\left(t\right) + K_I \int e\left(t\right) dt + K_D \frac{de(t)}{dt}. \tag{3.1}$$

As shown in Equation (3.1), the PID control strategy is triggered by applying k_P, k_I, and k_D on corresponding errors $e\left(t\right)$, which is simple and easy to realize, as shown in Fig. 3.2.

In trajectory tracking, PID controller is usually used to control the steering angle to make the vehicle follow the desired path given by the path planner by measuring the lateral offset and the yaw rate. For the output of a PID controller depends on three specific parameters for each application, which in our case is minimizing the lateral offset and driving a vehicle smoothly track the desired trajectory [54]. The influence of proportional, integral, and differential terms on the control effect is shown below.

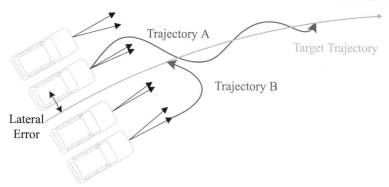

Figure 3.3: Influence of K_P on tracking performance.

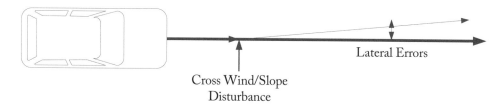

Figure 3.4: Influence of K_I on tracking performance.

(1) The Proportional Term

The proportional term produces a steering angle that is proportional to the lateral offset between vehicle and target trajectory, as shown in Fig. 3.3. The steering angle can be adjusted by multiplying the error by a constant K_P. If the proportional gain is too large, the system will become unstable, and the future movement of the vehicle is shown in the Trajectory B in Fig. 3.3. In contrast, if the proportional gain is too small, then the PID controller is not sensitive to the input error, and the same input error will result in a smaller output response. The advantage of this situation is that the system is stable in response to system disturbances. Tuning theory and industrial practice indicate that the proportional term should contribute to the bulk of the output change.

(2) The Integral Term

The contribution from the integral term $K_I \int e(t)dt$ is proportional to both the magnitude of the error and the duration of the error, as shown in Fig. 3.4. The integral term is the sum of the instantaneous error over time and gives the accumulated offset that should have been corrected previously. The accumulated error is then multiplied by the integral gain k_I and added to the controller output.

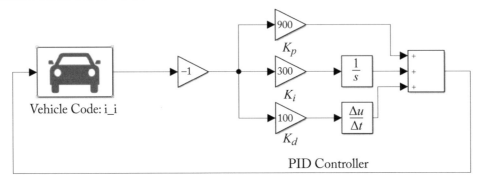

Figure 3.5: Block diagram of PID controller for path tracking.

(3) The Derivative Term

When the tracking error is small and gradually decreases, too big steering angle applied to the vehicle will lead to overshoot. Term $K_D \frac{de(t)}{dt}$ aims at flattening the error trajectory into a horizontal line and reducing overshoot. Derivative term is calculated by determining the slope of the error over time $\frac{de(t)}{dt}$ and multiplying this rate of change by the derivative gain K_D, the more rapid the change, the greater the controlling or dampening effect. Derivative term can predict vehicle's behavior, thus improve settling time and stability of the path-tracking system, and it is a best estimate of the future trend of the tracking error.

A simple PID controller is adopted in the co-simulation for path-tracking, and the simulation block diagram is shown in Fig. 3.5.

3.2   PREVIEW-FOLLOWING CONTROL

In situations where an experienced driver controls the vehicle, the vehicle follows the expected trajectory ahead accurately and smoothly. Therefore, it is considered to be a feasible way to design a human-like controller for the path-tracking task of autonomous vehicle [55]. From the perspective of the hypothesized driver behavior, the path-tracking control can be divided into two steps: the driver's early perception and steering operation, that is, the driver senses the vehicle attitude as well as the lateral displacement in advance and carries out a suitable steer to follow the expected trajectory.

Depending on the purpose of path-tracking, different methods can be used to simulate the driver's behavior. No matter which approach is adopted, the driver's steering operation process should be accurately understood. Assume that the driver inside the vehicle looks forward to a point in front of the vehicle and estimates the lateral displacement deviation of the vehicle relative to the target trajectory [56]. The estimated lateral displacement will serve as the decision basis of the driver's steering and the input signal of the feedback controller. If the system is

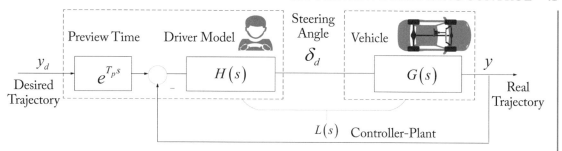

Figure 3.6: A closed-loop preview-following driver model.

defined as a closed control loop as illustrated in Fig. 3.6, methods from control theory may be applied.

In Fig. 3.6, the human driver model is considered as a controller $H(s)$, the vehicle is considered as a plant $G(s)$, and $L(s)$ is the controller-plant combination, T_p is the preview time or the time lag to make an action, δ_d is the output of the controller which is the steering angle of front wheel, y and y_d are actual and desired lateral positions, respectively. Since the driver's operation during normal driving is a low-frequency process, the transfer function of the closed-loop system should always be 1 at the low frequency range.

The preview-following method describes the tracking properties in accordance with the future input information of planned trajectory. Therefore, the preview time is a core concept in most driver models and an important parameter to determine the performance of the path-tracking controller. As the vehicle travels on a curved trajectory, the driver tends to adjust his or her preview time, which is usually between 0.5 and 2 sec, according to the curvature of the road ahead. The preview-following model shown in Fig. 3.7 can be used to express and simulate the vehicle state of motion and driver tracking behavior. If the X-axis is the direction along the predetermined trajectory, and the Y-axis is the direction perpendicular to it.

In 1953, Kondo built a classic preview driver model, and many subsequent models borrowed the ideas from this model in the design process [57]. This driver model has been widely applied because of its simple structure and clear physical meaning. In the case of preview-following control, it is assumed that the driver focuses on the preview point in front of the vehicle and estimates the lateral error between the vehicle and the trajectory based on the preview information. Depending on the number of preview points selected, preview models can be divided into single-point and multiple-points model, each defined by a preview time T_p according to

$$L_p(t) = V(t)T_p, \tag{3.2}$$

where $V(t)$ is the longitudinal speed and $L_p(t)$ is a varying distance to the point in front of the vehicle. The distance $L_p(t)$ between its current and preview points is really important for the formulation of this approach. If the look ahead distance is too short, the vehicle motion oscillates and becomes unstable. If $L_p(t)$ is big, the vehicle is stable and could return to the course. For

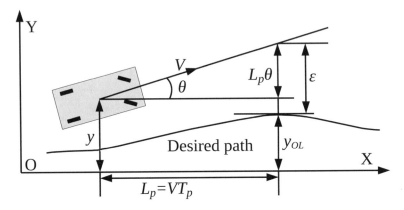

Figure 3.7: Preview-following control for path tracking.

the vehicle motion to be stable enough, $L_p(t)$ must increase with speed. In the controller design of path-tracking, it is preferable to ensure a small look-ahead distance in order to enable the vehicle to track the path smoothly.

Most driver models usually use deviations in lateral position or/and yaw angle to calculate the steering angle suitable for path-tracking. A simple example of a preview driver model that only uses the heading error is given be

$$\delta(t) = K\epsilon(t - T_R),\tag{3.3}$$

where $\epsilon(t)$ is the angle between the vehicle heading and the preview point, T_R is the driver reaction time, and K is a gain constant.

In addition, under the driving conditions shown in Fig. 3.7, y is the vehicle lateral displacement, θ is the yaw angle, and y_{OL} is the lateral displacement of the target course at the look ahead point. Since $|\theta| \ll 1$, the lateral deviation $\epsilon(t)$ of the vehicle from the desired trajectory at the look ahead point is:

$$\epsilon(t) = y + L_p\theta - y_{OL}.\tag{3.4}$$

Unlike the single-point preview model, the multiple-points model assumes that drivers focus on a front regional road to get multiple points of deviation. Generally, parameter optimization method is used to make the tracking performance of the multiple points model better, but also it makes the physical concepts of the model unclear.

For the preview-following model, the vehicle motion is not only dependent on the value of preview distance and the steering angle obtained, but also affected by the vehicle's inherent dynamics, and by the human driver's control characteristics.

In the design of the path-tracking controller, the driver's preview and following characteristics will make the tracking performance of the autonomous vehicle more consistent with reality. However, there is still no certain guarantee that the human driving operation can be

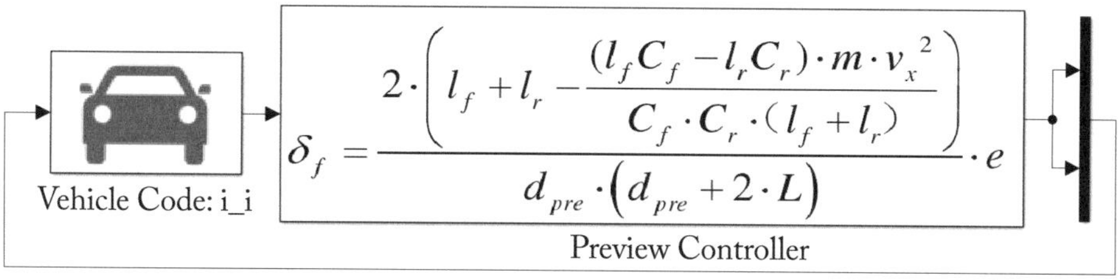

$$\delta_f = \frac{2 \cdot \left(l_f + l_r - \dfrac{(l_f C_f - l_r C_r) \cdot m \cdot v_x^2}{C_f \cdot C_r \cdot (l_f + l_r)} \right)}{d_{pre} \cdot \left(d_{pre} + 2 \cdot L \right)} \cdot e$$

Preview Controller

Figure 3.8: Vehicle motion controlled by the driver model.

described completely using mathematical equations. Therefore, it is necessary to identify and design the parameters of the preview control system according to specific application scenarios.

A simple preview-following controller is adopted in the co-simulation for path-tracking, and the simulation block diagram is shown in Fig. 3.8.

3.3   MODEL PREDICTIVE CONTROL

Model predictive control (MPC) is a practical control technology that can systematically considers the future predictions of plant response and system operating constraints in design stage. That makes it a suitable choice for path-tracking where the system faces dynamically traffic environment and has to satisfy crucial safety constraints (i.e., collision avoidance) as well as physical constraints (i.e., actuator saturation) [58].

In the path-tracking application of model predictive control algorithm, the control signal of vehicle actuators is usually obtained by solving the optimization problem, which is usually composed of an optimization objective of safety and tracking performance as well as the constraints of both the control actions and the plant outputs [59]. A vehicle model is used to predict the future motion state and absolute position of the vehicle under the current control input. Based on this prediction, at each time step t, a performance index is optimized under operating constraints with respect to a sequence of future input moves. The first of such optimal moves is the control action applied to the plant at time t. At time $t + 1$, a new optimization is solved over a shifted prediction horizon, as shown in Fig. 3.9.

As mentioned above, a unique feature of MPC-based path-tracking method is that it continuously optimizes performance indexes by receiving information of vehicle position, heading angle, and obstacles as the vehicle moves. Therefore, how to determine the appropriate performance index and select a reasonable optimization solver according to the task of path-tracking is the key to ensure the real time and accuracy of the controller. Figure 3.10 is the block diagram of model predictive control for path-tracking, in which the controller can calculate the future input sequence of the front steering angle for collision avoidance in a defined horizon.

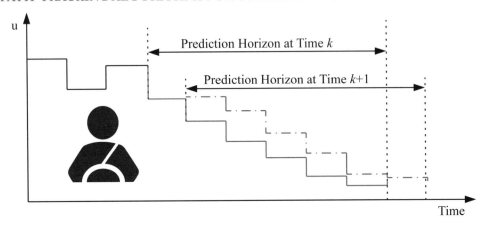

Figure 3.9: Iteration of model predictive control.

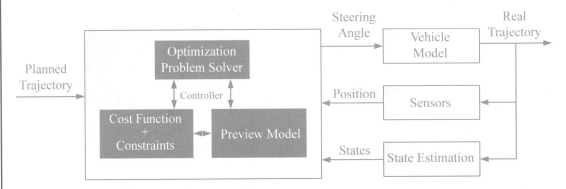

Figure 3.10: Model preview control for path tracking.

Because MPC is implemented with digit computers in most cases, the discrete error dynamics models are usually used in most MPC applications to predict the state of our vehicle at time $t + 1$ from the last state at time t, and the errors in the model are mainly lateral distance error and heading angle error. Using the kinematic bicycle model, we can easily deduce the location, the heading angle and the speed from the last state. The discrete-time vehicle model can be represented by

$$x_{k+1} = f(x_k, u_k)$$
$$y_k = g(x_k).$$

(3.5)

The states x_k, controls u_k, and outputs y_k are vectors.

MPC is an optimal-control method to calculate control inputs by minimizing an objective function. The design of the objective function is a critical step and can be defined in terms of ex-

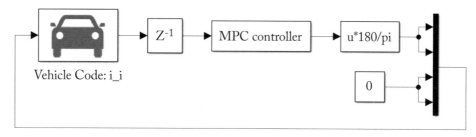

Figure 3.11: Design of model predictive controller for path tracking.

isting and predicted system variables. An MPC controller uses the following scalar performance measure for output reference tracking:

$$\min_{U_t} J_N\left(\overline{\xi_t}, U_t\right) = \sum_{k=t}^{t+N-1} cost\left(\xi_{k,t}, u_{k,t}\right) \tag{3.6a}$$

$$\text{Subj. to} \quad \begin{aligned} \xi_{k+1,t} &= f\left(\xi_{k,t}, u_{k,t}\right) \quad k = t, \dots, t+N-1 \\ \xi_{k,t} &\in \chi \quad k = t, \dots, t+N-1 \\ u_{k,t} &\in U \quad k = t, \dots, t+N-1, \end{aligned} \tag{3.6b}$$

where the symbol $x_{k,t}$ stands for the variable x at time k predicted at time t and N is the prediction horizon. ξ is the state of the system and in path-tracking includes the lateral position and yaw angle. u is the control input to the system and in this case includes the steering angle, braking torque, or throttle. The function $f(\xi, u)$ allows to predict the future system states based on the current states and control inputs. The second and third term in Equation (3.6b) are the state and input constraints the controller has to respect. The cost in Equation (3.6a) can be any performance index and can include penalties on state tracking error, penalties on inputs, and penalties on input change rate, etc.

According to the mentioned basic principles of MPC, the co-simulation block diagram of MPC application in path-tracking is established, as shown in Fig. 3.11.

MPC can be applied to under-actuated or over-actuated systems, this is really helpful in the case of collision avoidance where the actuators intervention is crucial in the path-tracking actions. Besides, the cost function containing system constraints provide an optimization objective to be solved in each sampling period. This in return ensures reliable control actions for collision avoidance and gives the MPC an advantage over other control technologies. Another advantage of MPC over other types of controllers is its intuitive tuning options. This includes the future prediction of the manipulative input, control horizons as well as the constraints concept that can manipulate the weights on the slip angle, steering angle, and lateral position.

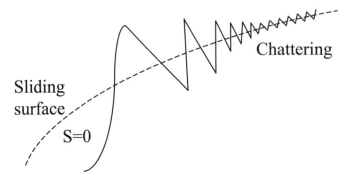

Figure 3.12: Sliding surface and effect of chattering in sliding mode controllers.

However, for a complex collision avoidance situation, the basic MPC control strategy needs to be improved. This can be done by utilizing multiple objective functions, reference trajectory as well as the need to use Nonlinear Model Predictive Control (NMPC). Owing to the presence of better computational devices, more complex MPC formulations which were infeasible in the past are now possible.

3.4 SLIDING MODE CONTROL

There are many nonlinear components in real vehicles, such as nonlinear tires and nonlinear suspensions, but it is difficult for the traditional linear control method to fully consider these nonlinear factors when building the vehicle model, which will directly affect the overall performance of path-tracking control system [60]. As another classical control strategy, sliding mode control has also been widely used in path-tracking applications due to its nonlinear characteristics. In sliding mode control, the state feedbacks as well as the control signals are treated as discontinuous functions which make it unaffected to parametric uncertainties and external disturbances. This is the main reason why the sliding mode controller is more suitable for path-tracking.

Sliding mode controllers have been successfully employed in many researches on path-tracking control for mobile robots as well as vehicle platforms. The controller works by applying a discontinuous control signal to force the system to move along the predetermined state trajectory of sliding mode. In most studies, sliding mode refers to the motion of the system along the sliding surface of the control structure [61]. The control law utilizes fast switching strategy in order to drive and maintain the system's state trajectory towards the chosen sliding surfaces as depicted in Fig. 3.12. Unlike other control strategies, the structure of control system is not fixed, which changes during the dynamic process according to the current state of the system. Therefore, its main advantage is that it does not simplify the dynamics through linearization and it has the advantages of rapid response, insensitive to changes in corresponding disturbances.

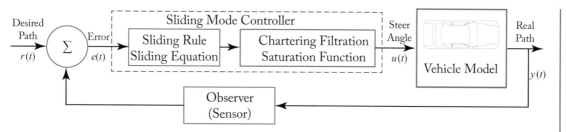

Figure 3.13: Sliding mode controller for path tracking.

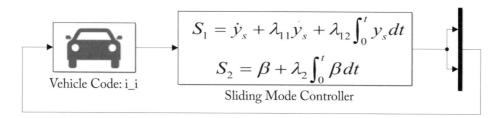

Figure 3.14: Co-simulation block diagram of sliding mode control.

The structure block diagram of the sliding model controller applied to path-tracking is shown in Fig. 3.13. The sliding mode controller for path-tracking can be designed by using MATLAB and CarSim software, as shown in Fig. 3.14.

However, due to the fast switching solution offered by this controller, the chattering of the control signal is the main drawback of the sliding mode control. With the delay and imperfections in physical actuators, it may lead to serious actuator and plant damages, energy losses, and unwanted disturbance due to these chattering. To reduce the system chattering, several methods have been proposed, such as high-order SMC, fast terminal SMC, and SMC with neural work algorithm.

3.5   FUZZY LOGIC CONTROL

In recent years, intelligent control algorithms have been increasingly applied to the autonomous driving system of vehicles, including fuzzy logic, neural networks, and genetic algorithms. Among them, fuzzy logic is a biological heuristic algorithm that controls the vehicle to realize path-tracking by imitating the human brain judgment and reasoning mode of think [62]. Similar to the preview-following control method, it is also an excellent path-tracking algorithm to simulate the driver's judgment and control behavior.

Fuzzy logic controller has ability to deal with numerical and linguistic problems simultaneously. The key step of establishing fuzzy control strategy is to design reasonable fuzzy sets and fuzzy logic according to the research object [63]. Fuzzy sets represent linguistic terms such

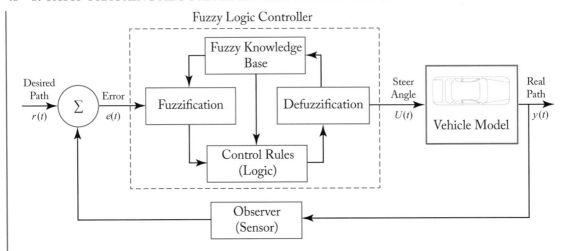

Figure 3.15: Fuzzy logic controller for path tracking.

as small, medium, large, and so on, while fuzzy logic can distinguish the fuzzy sets and deal with fuzzy relations with the help of membership function concept. Fuzzy logic is good at expressing qualitative knowledge and experience with unclear boundaries, which can be regarded as the extension of the rule-based controllers. With the linguistic description, the fuzzy control strategy has the advantage of its strong fault-tolerant ability and robustness to the change of dynamic characteristics, environmental characteristics, and moving conditions of the control object. Therefore, the fuzzy control strategy is suitable for uncertain, time-varying, strong nonlinear, and time-delay system, such as vehicle dynamic model. In addition, because fuzzy logic method has the ability to guide the controller to make decisions, many researches combine it with other controllers, such as fuzzy PID control, fuzzy sliding-mode control, and neuro-fuzzy control.

Fuzzy logic control has been used to solve the path-tracking problem of collision avoidance in many researches [64]. The basic structure of the fuzzy logic controller is usually composed of three blocks: the fuzzification, inference, and defuzzification, as shown in Fig. 3.15.

The first step to realize a fuzzy controller is fuzzification, which transforms each real value of inputs (such as lateral distance error) and outputs (such as the steering angle of front wheel) into grades of membership for fuzzy control terms.

The second part is fuzzy inference, which combines the facts obtained from the fuzzy rule base to carry out the fuzzy reasoning process. Different methods of fuzzy inference can be adopted according to the use and form of membership function (such as triangular, trapezoidal, gaussian, and so forth). Once the inputs, the outputs, and membership function are defined, the fuzzy rules can be determined based on the driving habits of experienced drivers.

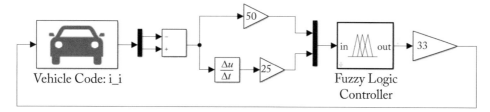

Figure 3.16: Co-simulation of path-tracking using fuzzy logic controller.

The third part of the fuzzy logic controller is defuzzification block. The objective of this part is to transform the subsets of the outputs which are calculated by the inference engine. In this case, the output of the fuzzy controller is the front-wheel steering angles acting on the vehicle that are smooth and mutation-free.

Based on the above three steps, the fuzzy logic controller for path-tracking can be designed by using MATLAB and CarSim software, as shown in Fig. 3.16.

However, the performance of the fuzzy controller designed by different researchers will vary greatly, because the formulation of fuzzy control rules depends on the researchers' in-depth understanding of driving behavior. However, it is difficult to objectively judge the merits and demerits of different driving behaviors by quantitative methods. Therefore, how to effectively avoid the influence of human factors in the design of fuzzy logic controller of path-tracking still needs further research.

3.6   SUMMARY

A good path-tracking algorithm is the one that can allow the vehicle to follow the predetermined path for obstacle avoidance closely. At the same time, the algorithm should also be robust, that is, it can deal with the disturbance caused by the system's nonlinearity and environmental uncertainty. The main scope of this section is to review the control strategies for path-tracking, and several commonly used control algorithms are introduced in some detail. Traditional control method (i.e., PID, sliding model control), human-like control strategies (i.e., preview-following theory, model preview control), and the intelligent approach (i.e., fuzzy logic control) are discussed to solve the path-tracking problem and ensure the vehicle successfully avoids the collisions.

In order to compare the tracking performance of different control methods, this chapter designs an expected path for active collision avoidance using the quintic polynomial equation proposed in Chapter 2, and selects the same vehicle model in the CarSim software. The co-simulation models with the same tracking target and different control methods are established, as shown in Figs. 3.5, 3.8, 3.11, 3.14, and 3.16. Simulation results are shown in Fig. 3.17 indicated that all the designed controllers can achieve the goal of path-tracking, and each algorithm shows its own advantages and disadvantages, although the parameters of each control method in these

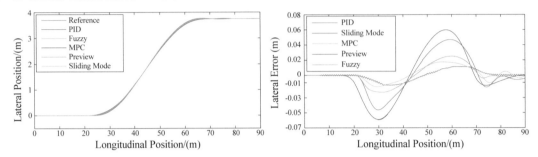

Figure 3.17: Simulation results of different control methods.

simulations are not optimal values. The selection of control strategies should be based on the requirement of the system and properties of the state variables that need to be controlled.

As evident from the review, any single-control algorithm has its limitations, and it is difficult to meet all the performance requirements of path-tracking task. A hybrid control scheme can be established by combining two or more control algorithms, which has a combined advantage of robustness, adaptability, rapid response, disturbance rejection, etc.; it is a promising approach to solve the more complex problem of path-tracking. However, such a hybrid system still needs to find a compromise according to the different control objectives, and determine which one is dominant in the system through system parameter adjustment, so as to guarantee its good performance in path-tracking.

CHAPTER 4

Optimal Local Trajectory for Vehicle Collision Avoidance Maneuvers

In this chapter, a path-planning method that generates a trajectory to mitigate the crash as much as possible is proposed for autonomous vehicles in emergency situations where an accident is unavoidable. When the avoidance of a collision is impossible for path-planning system, then the artificial potential field is filled into the controller objective to achieve the lowest possible severity. The MPC algorithm is adopted here for path-planning, and the vehicle dynamic is also treated as an optimal control problem. On account of the analysis above, the MPC can optimize the command following, obstacles avoidance, vehicle dynamics, road regulation, and mitigate the inevitable collision based on the predicted values. Simulations in this chapter have proved that the proposed MPC algorithm has the ability to avoid obstacles and mitigate the collision if the accident is inevitable.

4.1 INTRODUCTION

As mentioned in Chapter 1, a traffic accident is one of the main reasons for human casualties with statistics showing that millions of humans have lost their lives on the road annually. Determining how to generate a path with the lowest crash severity during inevitable accidents is an extremely prerequisite problem that must be addressed [65]. This technology challenge noted that reducing the level of harm until it is completely prevented is the objective of automated and connected driving [66].

For instance, Fig. 4.1 depicts an emergency for the silver ego vehicle, where obstacle vehicle 1 (orange color) is changing into the right lane. The ego vehicle is in its blind spot while obstacle vehicle 2 (bronze color) is driving in front of vehicle 1 at a slower speed. In this situation, there are two options for the ego vehicle, shown as the green and blue dotted lines in this figure. If the ego vehicle chooses the green line, it will experience a full-frontal collision with obstacle vehicle 1 at maximum deceleration. Alternatively, if the blue line is selected, the ego vehicle will crash into obstacle vehicle 2 at a certain angle. Which path should the ego vehicle select that will result in the least severe crash? That is the main attention of this research—to

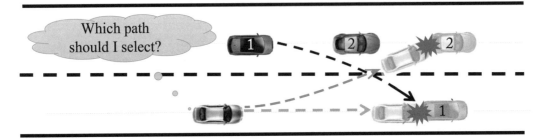

Figure 4.1: How can you generate a path that can mitigate the crash if a collision is unavoidable? If the ego vehicle chooses the green line, it will experience a full-frontal collision with the obstacle vehicle 1 at maximum deceleration; if the blue line is selected, the ego vehicle will crash into obstacle vehicle 2 at an angle.

generate a path that can avoid obstacles and mitigate crash as much as possible if the collision is unavoidable.

A survey of road accidents suggested that the most common accident scenarios could be classified as: the collision with a static obstacle on the road; the front/rear collision with vehicle around; the collision with adjacent vehicle when changing lane emergently; the collision with another vehicle running in the opposite lane direction, and the collision with a pedestrian involved [67]. The severity is dependent upon the nature of the obstacle (pedestrian, vehicle, road boundary, etc.), the crash speed, and its configuration. For the vehicle to vehicle collision, the crash severity mainly depends on the crash speed, the collision direction, the vehicle mismatch, driver features such as gender, age, etc., size of cars, and vehicle safety devices. How to design a path which has the lowest crash cost when avoidance is impossible, combining these above crash severity factors, is an ongoing research topic.

Trajectory-based optimization approaches have become the state-of-the-art path-planning algorithms for autonomous vehicles in recent years. The core of this technique is formulating the path-planning problem as an optimization problem, taking into account the desired vehicle performance and multi-relevant constraints [68]. MPC has been proven well suited for solving the path-planning problem, because of their ability to handle multi-constraints and convex problems. MPC solves a sequence of finite time trajectory optimization problems in a recursive manner and can think about the updating of the environment states during its planning process. Therefore, we choose MPC to solve the path-planning problem in this chapter.

To our best knowledge, crash mitigation research is still a frontier issue in the area of path-planning. In this case, an MPC is implemented for motion-planning to generate a path which can avoid general obstacles and, when collision is inescapable, generates an emergency path with the lowest crash severity. In the MPC predicted crash severity, the artificial potential field of obstacles and road boundaries, path-following matrix, and other vehicle performance

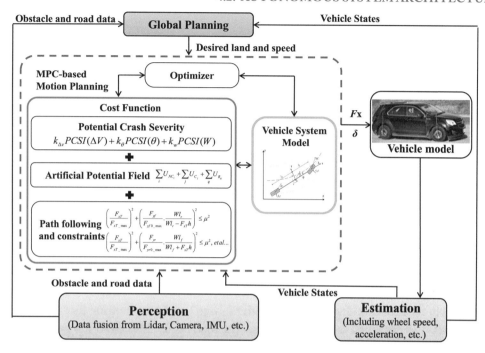

Figure 4.2: Autonomous system architecture.

constraints are considered in the cost function. Different scenarios are simulated to verify our proposed control strategy and its ability to generate the path to both avoid obstacles and mitigate the crash severity to provide the safest autonomous vehicle.

4.2   AUTONOMOUS SYSTEM ARCHITECTURE

A completely autonomous system includes modules for perception, global planning, motion planning, vehicle model, and estimation modules, as shown in Fig. 4.2.

The perception module is essentially the data fusion process. The data from different sensors including Lidar, camera, GPS, IMU, etc. are fused to obtain information regarding vehicle location, obstacles (nature, position, velocity, and shape), road structures, etc. [69]. The obtained information, combined with the vehicle's states calculated from estimation module and the global trajectory, is put into the MPC-based motion planning module. The goal of motion planning is to produce a path that follows the results of global-planning module while in accordance with the road regulations, having stable vehicle dynamics, and avoiding the obstacles, or having the lowest crash severity when avoidance is unachievable. The steering angle and the accelerator/brake pedal positions are calculated by the motion-planning module [70].

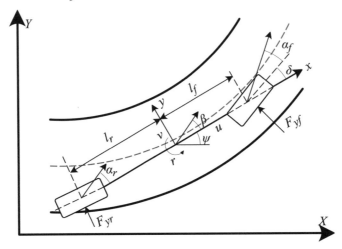

Figure 4.3: Vehicle bicycle model for path tracking.

The main focus of this chapter is on the MPC-based motion-planning for the autonomous driving system. The information of obstacles and the road boundaries from the perception module, the desired lane, and speed information from the global planning module, and the states of the vehicles from the estimation module are assumed to be received by the motion-planning module.

4.3 CRASH MITIGATION MOTION PLANNING

The control design for crash mitigation motion-planning is introduced in this section. The process includes modeling of the vehicle, the definition of the severity factor, the introduction of the artificial potential field, and the MPC algorithm for motion-planning.

4.3.1 VEHICLE DYNAMIC MODELING

The dynamic behaviors of vehicle are complicated in reality. There should be a trade-off between model accuracy and computational cost. To design the controller, a two degree-of-freedom (DOF) bicycle model and a 1-DOF longitudinal model are used with certain assumptions. The model is sufficient in the driving condition of high-speed and small-angle cornering, i.e., lane change in this chapter. Figure 4.3 depicts the vehicle model with the longitudinal, lateral, and yaw dynamics governed by:

$$m(\dot{u} - vr) = F_{xT} \tag{4.1}$$

$$m(\dot{v} + ur) = F_{yf} + F_{yr} \tag{4.2}$$

$$I_z \dot{r} = F_{yf} \, l_f - F_{yr} \, l_r. \tag{4.3}$$

The vehicle's motion with respect to global coordinates:

$$\dot{X} = u\cos\psi - v\sin\psi, \qquad \dot{Y} = u\sin\psi + v\cos\psi, \qquad (4.4)$$

where X and Y are the vehicle's longitudinal/lateral positions with respect to the global coordinates; ψ denotes vehicle heading angle; u and v represent vehicle longitudinal/lateral velocities, respectively; r represents the vehicle's yaw rate at its CG (center of gravity); m is the vehicle's mass; I_z is the vehicle's moment of inertia; l_r and l_f mark the distances from the vehicle CG to the rear/front axles, respectively; F_{xT} is the total tires longitudinal force; and F_{yf} and F_{yr} are the total lateral forces of the front/rear tires, respectively.

The vehicle is assumed to have a drive-by-wire front steering system. Utilizing a linear tire model, the lateral forces are developed as:

$$F_{yf} = -C_{\alpha f}\alpha_f = C_{\alpha f}\left(\delta - \frac{v + l_f r}{u}\right) \qquad (4.5)$$

$$F_{yr} = -C_{\alpha r}\alpha_r = C_{\alpha r}\left(-\frac{v - l_r r}{u}\right), \qquad (4.6)$$

where α_r and α_f denote the front and rear tires' sideslip angles, δ is the front steering angle, and $C_{\alpha f}$ and $C_{\alpha r}$ denote the cornering stiffness of the front and rear tires.

4.3.2 DEFINITION OF THE POTENTIAL CRASH SEVERITY INDEX (PCSI)

As presented in Section 4.1, the severity of an accident lie on a combination of factors, i.e, crash speed, the characteristics of the obstacle, and the hitting status (front collision with a rigidly-fixed obstacle, with another vehicle, or with a parked vehicle) [71]. Three core factors are thought over in this chapter: crash speed, crash angle, and mass ratio of two collision vehicles.

(1) Relative Speed ΔV

Several indexes associate with speed are adopted to evaluate the potential collision severity: the energy-equivalent speed (EES), the equivalent-barrier speed, the acceleration severity index (ASI), or occupant impact velocity (OIV). In this chapter, the relative velocity, ΔV, of the vehicle with respect to an obstacle (moving or static) is adopted as a metric to measure the potential crash severity index (PCSI) :

$$PCSI(\Delta V) = \frac{\Delta V}{D}, \qquad (4.7)$$

where ΔV and D denote the approaching velocity and the distance between ego vehicle and obstacle vehicle, respectively.

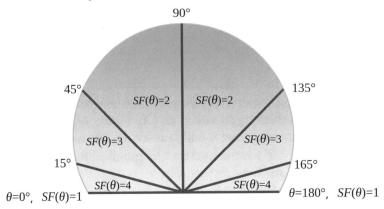

Figure 4.4: The potential severity factor related to the relative angle θ.

(2) Relative Heading Angle θ

From literature review and the accidents analysis, the crash angle according to the overlap of two vehicles in a collision can be divided into 3 categories: full overlap collision, 1/3 overlap near collision, and 2/3 overlap collision. The severity of these overlap collision categories was assessed from the data of 9902 accidents-involving Volvo vehicles, which demonstrated that the most severe risk happened around the 1/3 overlap collision, and when the equivalent barrier speed (EBS) is greater than 20 mph. 1/3 overlap crashes are 2 to 3 times higher compared to the corresponding risk in full-frontal crashes [72].

On account of the above analysis and for the convenience of implementation, we define the relative angle θ as the heading angle toward each other. The range of the relative heading angle is from 0–180°, and are split into 6 areas: 0–15°, 15–45°, 45–90°, 90–135°, 135–165°, and 165–180°. The potential crash severity index related to the relative angle θ is defined as described below:

$$PCSI(\theta) = \begin{cases} 1 & \theta = 0° \ \& \ 180° \\ 4 & \theta = 0° \sim 15° \ \& \ 165° \sim 180° \\ 3 & \theta = 15° \sim 45° \ \& \ 135° \sim 165° \\ 2 & \theta = 45° \sim 90° \ \& \ 90° \sim 135°. \end{cases} \tag{4.8}$$

(3) Mass Ratio Wo/W

In terms of mismatch in two-vehicle crashes, an analysis data by the Insurance Institute of Highway Safety (IIHS) reported that the relative risk of death for occupants in a passenger vehicle involved in frontal collisions when the light truck vehicle exceeds 3–4 times those involved in similar collisions with another passenger vehicle. The relative risk of death could reach 27–48 times higher for the occupants in the passenger vehicle for the side collision impacts [73]. Regarding the potential crash severity index related to the mass ratio of the two vehicles, we can simply formulate this as:

$$PCSI(W) = \frac{W_o}{W},$$ (4.9)

where W_o, and W are the weights of the obstacle vehicle and the ego vehicles, respectively.

To sum up, the total potential crash severity index is defined as follows:

$$PCSI = f(\Delta V, \theta, W_O/W) = k_{\Delta v}PCSI(\Delta V) + k_\theta PCSI(\theta) + k_w PCSI(W),$$ (4.10)

where $k_\Delta v$, k_θ, and k_w are the weights of potential crash severity index related to the relative speed, relative heading angle, and mass ratio, respectively.

4.3.3   OBSTACLE DESCRIPTION

An artificial potential field (PF) is a field consists of obstacles to guide the vehicle toward the destination, simultaneously keeping it away from the obstacles including the non-crossable, crossable, and road boundary. There are three kinds of obstacles defined by the artificial potential field, non-crossable obstacle like a car or pedestrian (UNC), crossable obstacle like a bump (UC), and the road (UR) markers or boundaries. The overall potential field is generated from the sum of obstacles PFs:

$$U = \sum_i U_{NC_i} + \sum_j U_{C_j} + \sum_q U_{R_q},$$ (4.11)

where subscripts i, j, and q represent the index of the non-crossable road obstacle, crossable road obstacle, as well as road lane marker, respectively.

A detailed introduction to these three PFs is provided next.

(1) Non-Crossable Obstacle
Non-crossable obstacles such as a vehicle or a pedestrian could lead to damage or instability to a vehicle, or even worse threatens people's lives. The PF of non-crossable obstacle is generated by using a hyperbolic function of the gap of the ego vehicle with respect to the obstacle, as a function of safe distance, s_i:

$$U_{NC_i}(X, Y) = \frac{a_i}{s_i^{b_i}} = \frac{a_i}{s_i \left(\frac{dX}{X_{si}}, \frac{dY}{Y_{si}} \right)^{b_i}}$$

$$X_{si} = X_0 + uT_0 + \frac{\Delta u_{a_i}^2}{2a_n}$$ (4.12)

$$Y_{si} = Y_0 + (u \sin \theta + u_{oi} \sin \theta)T_0 + \frac{\Delta v_{a_i}^2}{2a_n},$$

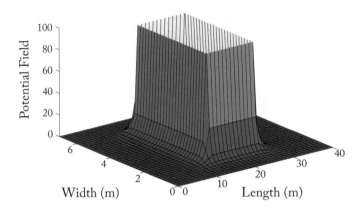

Figure 4.5: PF of the non-crossable obstacle.

where a_i and b_i are the shape and intensity parameters of the PF, respectively, dX and dY denote the distance between the obstacle and the ego vehicle in longitudinal and lateral direction, respectively. X_o and Y_o are the minimum longitudinal/lateral distances, Y_{si} and X_{si} are the safe lateral and longitudinal distances from the obstacle, respectively; T_0 denotes the safe time gap, u and u_{oi} are the velocity of the ego vehicle and obstacles, respectively; θ is the heading angle toward each other, and Δu_{ai} and Δv_{ai} are the longitudinal/lateral approaching velocities. By and large, the potential field of non-crossable obstacle is related to the distance between the obstacle and the ego vehicle, the shape and intensity parameters of the obstacles, the heading angle toward each other, and the approaching velocities.

The PF of a moving vehicle at 80 km/h located at $(X_{oi}, Y_{oi}) = (20, 3.5)$ is shown in Fig. 4.5.

(2) Crossable Obstacle
The PF of some obstacles such as a little bump or some soft garbage on the road cannot cause severe damage to the ego vehicle. The PF is defined with the following exponential function:

$$U_{C_j}(X, Y) = a_j e^{-b_j s_j} = a_j e^{-b_j s_j \left(\frac{dX}{X_{si}}, \frac{dY}{Y_{si}} \right)}, \tag{4.13}$$

where a_j and b_j are the shape and intensity parameters of the obstacle, s_j is the normalized safe distance between the obstacle and the ego vehicle which is calculated similarly to Equation (4.12). Similar to the non-crossable obstacles, the potential field of crossable obstacle relates with distance between the obstacle and the ego vehicle, the shape and intensity parameters of the objects, the heading angle toward each other, and the approaching velocities. Figure 4.6 shows the PF of a crossable obstacle located at 10 m, 2 m.

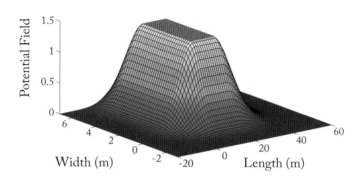

Figure 4.6: PF of crossable obstacle.

(3) Road Boundary

When the ego vehicle drives on the road, especially on the highway, the vehicle cannot depart from the road lane marker unless a lane change is required, in addition, hitting the road isolation belt is forbidden as it will cause instability or a serious crash accident. Quadratic form functions are utilized here to formulate the PFs of the road boundary in order to avoid undesirable road crossings:

$$U_{R_q}(X,Y) = \begin{cases} a_q\left(S_{Rq}(X,Y) - D_a\right)^2 & S_{Rq}(X,Y) \ < \ D_a \\ 0 & S_{Rq}(X,Y) \ > \ D_a, \end{cases} \tag{4.14}$$

where s_{Rq} is the safe distance of the vehicle from the road boundary, D_a is the permitted distance from road boundary, index $q = r, l$ is the right or left road lane marker, and a_q represents the intensity parameter. The conclusion can be drawn that the potential field of road boundary is related to distance between the ego vehicle and the road boundary, the permitted distance from road boundary, and the intensity parameter of the road boundary.

If remaining within a lane is intended, the right and left lane markers are the ones on which the PFs are implemented. If a lane change is intended, the PF is not implemented on the lane marker to be crossed for the lane change. It is implemented on the next lane marker instead. The PF of a two-lane road can be seen from Fig. 4.7, the further a vehicle deviates from the center of the lane, the harder the PF would push it back toward the lane center.

4.4   CONTROLLER DESIGN FOR MOTION PLANNING

The MPC algorithm is adopted for motion-planning in this section. To achieve general obstacles avoidances, and the lowest crash severity when the system determines that avoidance is impossible, the presented crash severity factor and artificial potential field are applied to the controller objective [74]. In addition, the vehicle dynamic is also regarded as an optimal control

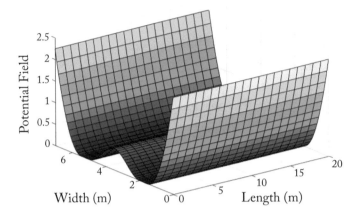

Figure 4.7: PF of a road with two lanes.

problem. Based on the analysis above, the MPC can optimize the command following, obstacle avoidance, vehicle dynamics, and mitigate the inevitable crash on the basis of the prediction. It can be assumed that the motion planning module receives information on the desired lane and speed; obstacles, road boundaries, and the vehicles states.

Synthesizing Equations (4.1)–(4.6), the vehicle's operating point dynamic in global coordinates can be written in state space form as:

$$\dot{x} = Ax + Bu \tag{4.15}$$

$$y = Cx, \tag{4.16}$$

where, $x = [X \ u \ Y \ v \ \psi \ r]^T$, $u = [F_{xT} \ \delta]^T$, $y = [Y \ u]^T$,

$$A = \begin{bmatrix} 0 & 1 & 0 & 0 & 0 & 0 \\ 0 & 0 & 0 & 0 & 0 & 0 \\ 0 & 0 & 0 & 1 & u & 0 \\ 0 & 0 & 0 & -\dfrac{C_{\alpha f} + C_{\alpha r}}{mu} & 0 & \dfrac{l_r C_{\alpha r} - l_f C_{\alpha f}}{mu} - u \\ 0 & 0 & 0 & 0 & 0 & 1 \\ 0 & 0 & 0 & \dfrac{l_r C_{\alpha r} - l_f C_{\alpha f}}{I_z u} & 0 & -\dfrac{l_f^2 C_{\alpha f} + l_r^2 C_{\alpha r}}{I_z u} \end{bmatrix},$$

$$B = \begin{bmatrix} 0 & \dfrac{1}{m} & 0 & 0 & 0 & 0 \\ 0 & 0 & 0 & \dfrac{C_{\alpha f}}{m} & 0 & \dfrac{l_f C_{\alpha f}}{I_z} \end{bmatrix}^T,$$

$$C = \begin{bmatrix} 0 & 0 & 1 & 0 & 0 & 0 \\ 0 & 1 & 0 & 0 & 0 & 0 \end{bmatrix}.$$

It should be noted that in Equation (4.15) the longitudinal dynamics are decoupled from the lateral and yaw dynamics, which means that the term vr in Equation (4.1) is ignored in formulating the state-space form. It makes sense as the term $\dot{u}$ dominates the longitudinal dynamics and the lateral speed v is (or should be) very small out of stability issue. For the following commands, the outputs to be tracked are the desired lateral position and the longitudinal velocity, defined as follows:

$$\begin{aligned} y_{des} &= \begin{bmatrix} Y_{des} & u_{des} \end{bmatrix}^T \\ Y_{des} &= (l_{des} - 1/2)L_w + \Delta Y_R, \end{aligned} \tag{4.17}$$

where y_{des} denotes the desired output matrix including the desired lateral position Y_{des} and desired vehicle speed u_{des}, L_w is the width of the lane, ΔY_R is the lateral offset of the road, and l_{des} denotes the index number for the desired lane.

One of the advantages of MPC is its ability to deal with multi-constraints. Constraints considered in this MPC problem include the road regulations, actuator capacity limits, and the constraints on the vehicle dynamics.

First, the vehicle should not violate the maximum and minimum speed limits specified by the road regulations. The limit is considered by a constraint:

$$u_{\min} < u < u_{\max}, \tag{4.18}$$

where $u_{\min}$ and $u_{\max}$ denote the minimum and maximum allowed speed.

The actuator capacities are considered as:

$$\begin{aligned} |\delta| &\leq \delta_{\max}, \\ F_{xT} &\leq \frac{T_{\max}}{R_{eff}}, \\ |\Delta\delta| &\leq \Delta\delta_{\max}, \end{aligned} \tag{4.19}$$

where $\delta_{\max}$ is the maximum steering angle, R_{eff} is the radius of the wheels, $T_{\max}$ is the maximum propelling torque, $\Delta\delta$ is the change of steering angle in one step, and $\Delta\delta_{\max}$ is its capacity.

Moreover, the MPC problem should consider tire force constraints because the lateral and longitudinal tire forces cannot exceed the friction ellipse:

$$\left(\frac{F_{xT}}{F_{xT\_\max}}\right)^2 + \left(\frac{F_{y*}}{F_{y*\_\max}}\right)^2 \leq \mu^2, \tag{4.20}$$

where $F_{xT\_\max}$ is the maximum tire longitudinal force, F_{y*}, for $* = f, r$, is the front or rear lateral tire force, $F_{y*\_\max}$, for $* = f, r$, is the maximum front or rear lateral tire force, and μ

is the tire-road friction coefficient. It is notable that these constraints also limit the lateral tire forces to remain in their linear region.

The maximum forces in the constraint equations of (4.20) are dependent on the load transfer. The lateral load transfer is ignored since the bicycle vehicle dynamics considers the total forces of the tires on the same wheel track. The effects of the longitudinal load transfers on the vertical forces of front and rear tire are:

$$F_{zf} = \frac{Wl_r - F_{xT}h}{l_f + l_r}$$
$$F_{zr} = \frac{Wl_f + F_{xT}h}{l_f + l_r},$$

(4.21)

where F_{z*}, for $* = f, r$, is the front or rear vertical tire force, W is the vehicle weight, and h is the height of the vehicle's center of gravity from the ground. However, the longitudinal load transfer affects the maximum lateral force in Equation (4.20). Assuming that the lateral tire force capacity changes linearly with respect to the vertical tire force:

$$F_{y*\_max} = F_{y*0\_max} \frac{F_{z*}}{F_{z*0}},$$

(4.22)

which F_{y*0_max} and F_{z*0}, for $* = f, r$, are the nominal maximum lateral front or rear tire force and the nominal vertical front or rear tire force, respectively, where nominal forces are the forces with no load transfer.

The longitudinal load transfer effect is included in the constraints of the tire force ellipse:

$$\left(\frac{F_{xT}}{F_{xT\_max}}\right)^2 + \left(\frac{F_{yf}}{F_{yf0\_max}} \cdot \frac{Wl_r}{Wl_r - F_{xT}h}\right)^2 \le \mu^2$$
$$\left(\frac{F_{xT}}{F_{xT\_max}}\right)^2 + \left(\frac{F_{yr}}{F_{yr0\_max}} \cdot \frac{Wl_f}{Wl_f + F_{xT}h}\right)^2 \le \mu^2,$$

(4.23)

where F_{yr0_max} and F_{yf0_max} are the nominal maximum lateral rear and front tire force. The constraints are nonlinear but convex. They are approximated by linear constraints so that they can be used in the quadratic MPC problem. Each constraint is approximated by a hexagon inscribed in it. The hexagon is calculated by minimizing the area between the original constraint and its hexagon approximation to maximize the available tire force for the vehicle model.

The cost function consists of the potential field U, severity factory $PCSI$, the tracking of the desired path, violation of the control input, and slack variables to penalize the violation. In this cost function shown in Equation (4.24), the tracking of the desired path, control inputs, violation of control inputs, and the slack variables are weighted by matrices Q, R, S, and P, respectively:

$$\min J = \min_{u_c, \varepsilon} \sum_{k=1}^{N_p} U^{t+k,t} + PCSI^{t+k,t} + \left\| y^{t+k,t} - y^{t+k,t}_{des} \right\|_Q^2 + \left\| u_c^{t+k-1,t} \right\|_R^2$$
$$+ \left\| u_c^{t+k-1,t} - u_c^{t+k-2,t} \right\|_S^2 + \left\| \varepsilon_k \right\|_P^2 \tag{4.24}$$

$$s.t. \left(k = 1, \ldots, N_p \right)$$

$$x^{t+k,t} = \mathbf{A}_d x^{t+k-1,t} + \mathbf{B}_d u_c^{t+k-1,t} \tag{4.0a}$$

$$y^{t+k,t} = \mathbf{C} x^{t+k,t} + \mathbf{D} u_c^{t+k,t} \tag{4.0b}$$

$$y_s^{t+k,t} = \mathbf{C}_s x^{t+k,t} + \mathbf{D}_s u_c^{t+k,t} \tag{4.0c}$$

$$y_s^{t+k,t} \le y_{s\_\max}^{t+k,t} + \varepsilon \tag{4.0d}$$

$$\varepsilon_k \ge 0 \tag{4.0e}$$

$$\mathbf{u}_{c\_\min} < \mathbf{u}_c^{t+k-1,t} < \mathbf{u}_{c\_\max} \tag{4.0f}$$

$$\Delta\mathbf{u}_{c\_\min} < \mathbf{u}_c^{t+k,t} - \mathbf{u}_c^{t+k-1,t} < \Delta\mathbf{u}_{c\_\max} \tag{4.0g}$$

$$u_c^{t+k,t} = u_c^{t+k-1,t}, k \ge N_c, k \ne c_2 N_{rc} + N_c, c_2 = 1, \ldots, (N_p - N_c)/N_{rc} \tag{4.0h}$$

$$u_c^{t-1,t} = u_c(t-1) \tag{4.0i}$$

$$x^{t,t} = x(t), \tag{4.0j}$$

where N_p is the prediction horizon; $t + k, t$ represents the predicted value at k steps ahead of t; ε_k denotes the slack variables vector at k steps ahead, which represents the penalty of soft constraints on tire forces; and N_c is the control horizon. The vehicle's states and tracking outputs are predicted through Equations (4.0a) and (4.0b), respectively, using the output and feedforward matrices denoted by $\mathbf{C}$ and $\mathbf{D}$. The constraints variables y_s are linearized as a function of the inputs as well as states in Equation (4.0c) where $\mathbf{C}_s$ and $\mathbf{D}_s$ denote the output and feedforward matrices, respectively. The constraints on the actuators, the vehicle's speed and corresponding linear constraints of the tire capacity constraints are presented in Equation (4.0d), where y_s is the soft constraint variables denoting the violation allowance of the bounds. The slack variables corresponding to actuator constraints are set to zero since they cannot be violated. Equations (4.0f) and (4.0g) denote the vehicle speed and its violation limitation, respectively. The number of control inputs is reduced in Equation (4.0h) which may save the computational cost. The proposed algorithm is solved for any PF. The problem is, however, nonlinear and non-convex since PFs are nonlinear as well as non-convex. So its solution is time consuming. Here, it is converted into its approximated quadratic convex problem to reduce the calculation time. For brevity, the details of the convex process are not shown here.

After the convex process, the optimal control problem becomes a convex quadratic optimization problem. The problem like a corresponding nonlinear problem solved by Sequential

Table 4.1: Parameters of autonomous vehicle and the controller

Parameter	Value	Unit	Parameter	Value	Unit
m	2,270	kg	a_{max}	8.8	m/s^2
h	0.647	m	a_n	1	m/s^2
l_f	1.421	m	D_a	0.5	m
l_r	1.434	m	N_p	20	--
I_z	4,600	kgm^2	N_c	5	--
C_f	127,000	N	N_{rc}	5	--
C_r	130,000	N	Q	[0.2 0.01]	
μ	0.9	--	R	[2e-9 100]	
R_{eff}	0.351	m	X	[5e-8 500]	
F_{xt_max}	20,000	N	δ_{max}	10	deg
F_{yf0_max}	10,400	N	$\Delta \delta_{max}$	0.5	deg
F_{yr0_max}	10,600	N	T_{max}	3000	Nm

Quadratic Programming (SQP) in one sequence. The specific solution of this problem is to derive an upper bound for the optimization error of each sequence of the SQP, where the optimization error is the difference between the result of the sequence and the local minimum of the nonlinear problem in the neighborhood of the problem's initial value. Considering this upper bound, for the quadratic problem, the closer the problem's initial value is to the minimum, which is equivalent to the anticipated vehicle point being closer to the vehicle position at the minimum, the smaller the optimization error. In the next section, the performance of this solver will be presented.

4.5   CASE STUDY

To verify the feasibility of the proposed crash mitigation algorithm, three scenarios are simulated and analyzed. Table 4.1 summarizes the parameters of the autonomous vehicle and the controller.

In Table 4.2, u_0 denotes the ego vehicle's initial velocity; u_{des} denotes the desired velocity of the ego vehicle; X_0 denotes the initial longitudinal position of the ego vehicle; V_{o1}, V_{o2} represent the initial longitudinal velocity of obstacles, respectively; and X_{o1}, X_{o2} represent the initial longitudinal position of obstacles, respectively.

Table 4.2: Initial information for test scenarios

	u_0 (km/h)	u_{des} (km/h)	X_0 (m)	V_{o1} (km/h)	V_{o2} (km/h)	V_{o3} (km/h)	X_{o1} (m)	X_{o2} (m)
Scenarios 1–2	80	80	0	0	–	–	–	80
Scenario 3	100	100	0	60	45	–	22	30

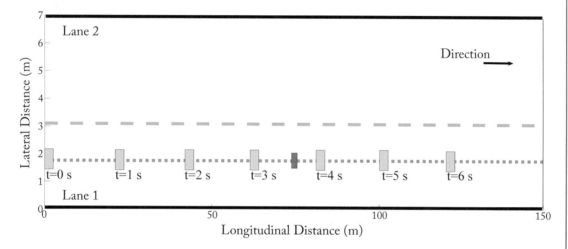

Figure 4.8: Positions of ego vehicle (yellow) and obstacle (red).

4.5.1 EVALUATION OF COLLISION AVOIDANCE

One qualified motion-planning algorithm should have the ability of collision avoidance. Two scenarios are adopted here to verify the collision avoidance ability of the proposed motion-planning algorithm.

Scenario 1: The ego vehicle is running on lane 1 and there is no lane change command. A static crossable obstacle is placed at 80 m ahead of the ego vehicle and 1.5 m from the right lane boundary. This crossable obstacle is marked by a 0.5 m purple square. The initial information is presented in Table 4.2. The trajectory of the ego vehicle (yellow) is demonstrated in Fig. 4.8 and shows the ego vehicle traveling straight without any lane change, because this obstacle is detected as a crossable obstacle. The longitudinal velocity and longitudinal force of the ego vehicle is shown in Fig. 4.9. We can find that the longitudinal velocity (orange) and force (green) are almost unchanged.

Scenario 2: This scenario is adopted to verify the non-crossable obstacle avoidance capability. The initial information is presented in Table 4.2. The ego vehicle is running on the lane 1,

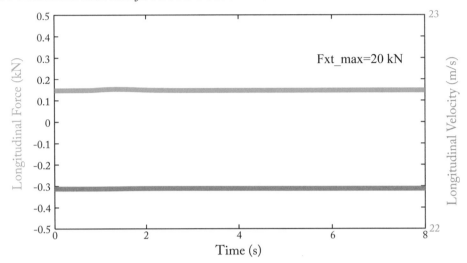

Figure 4.9: Longitudinal force (orange) and velocity (green) of the ego vehicle.

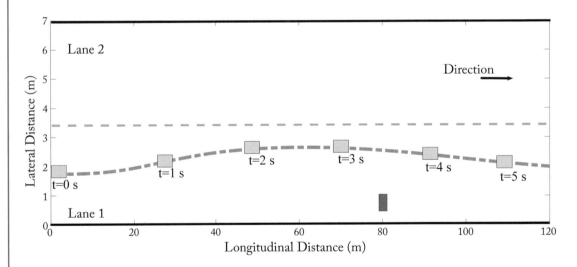

Figure 4.10: Positions of ego vehicle (yellow) and obstacle (red).

at the same time, a static non-crossable obstacle placed at 80 m ahead of the ego vehicle which is 0.5 m from the right lane boundary. This non-crossable obstacle is represented by a 0.5 m red square. In this situation, there is sufficient lateral space for the vehicle to pass it alone the same lane. The action of ego vehicle (yellow) and the position of the non-crossable obstacle (red) are depicted in Fig. 4.10. The vehicle avoids the non-crossable obstacle successfully and then comes

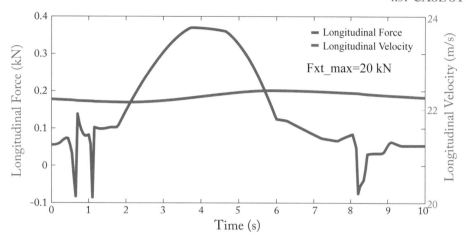

Figure 4.11: Longitudinal force and velocity of the ego vehicle.

back to the original lane 1, tracking the global trajectory. Figure 4.11 shows the longitudinal velocity (orange) and force (green) of the ego vehicle. The steering angle command directly from the wheel of the ego vehicle is shown in Fig. 4.12.

From Fig. 4.10, we can observe that the ego vehicle avoids the non-crossable obstacle smoothly and then comes back to track the global trajectory. We also can conclude from Figs. 4.11 and 4.12 that the action of the ego vehicle changes a little bit and relatively smoothly. Furthermore, the longitudinal force and steering angle command from the wheel are under upper limits.

Based on the simulation results and quantitative analysis of these above two scenarios, the brief conclusion can be drawn that both the crossable and non-crossable obstacles can be dealt with properly. Nevertheless, sometimes the obstacle or crash may not be avoided in a timely manner. Therefore, the two scenarios below are represented to prove the ability to mitigate the crash severity.

4.5.2 MITIGATE CRASH SEVERITY RELATED TO CRASH ANGLE

Scenario 3: The ego vehicle is running in lane 1 at 100 km/h. There is an obstacle, vehicle 1, on the same longitudinal position 15 m ahead in lane 2. This obstacle vehicle takes the lane change from lane 2 to lane 1 with a constant speed 60 km/h within $t = [1\ 2]$ s. There is another obstacle, vehicle 2, on lane 2 with the speed of 45 km/h. The situation is designed such that the ego vehicle cannot avoid both obstacles while it stays within the road boundaries as shown in Fig. 4.13 same as the situation introduced in Fig. 4.1.

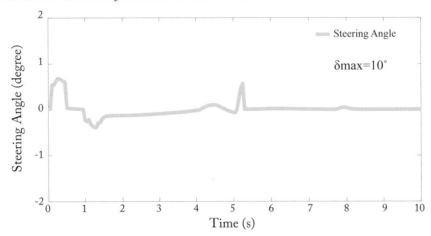

Figure 4.12: Steering angle command from the wheels of the ego vehicle.

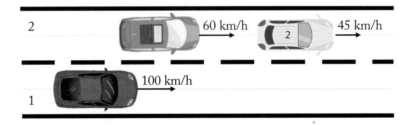

Figure 4.13: Schematic diagram of scenario 3.

There are two options for the ego vehicle: (1) trigger the autonomous emergency braking system while continuing to move forward, however this will result in rear impact from the approaching vehicle 1; and (2) use the emergency brake while turning left, but this will lead to a frontal angle collision with the following vehicle 2. The results of the motion-planning algorithm with and without crash mitigation are demonstrated as follows.

(1) Without Crash Mitigation
As shown in Fig. 4.14, the ego vehicle with green color selects the light green path, which crashes with obstacle vehicle 2 at a crash angle. The trajectory is marked every 1 s. The collision happens at 2.05 s when ego vehicle is (51.84 m, 3.25 m) while the obstacle vehicle 2 is (55.63 m, 5.25 m). The emergency brake starts at 1 s. This crash with angle may cause rollover and secondary crashes with other vehicles. The severity is much higher than the full front crash. The longitudinal force and velocity of the ego vehicle before the crash are shown in Fig. 3.15, and the steering angle command from the wheels before the crash is shown in Fig. 4.16. From Figs. 4.15 and 4.16,

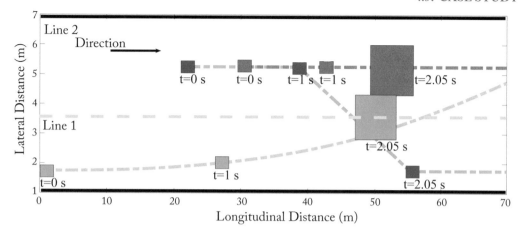

Figure 4.14: Positions of ego vehicle (green), approaching vehicle 1 (blue), and the obstacle vehicle 2 (red).

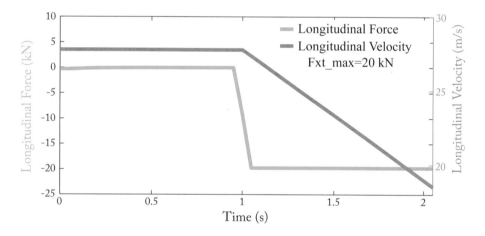

Figure 4.15: Longitudinal force and velocity of the ego vehicle before crash.

we can see that the action of the ego vehicle within the maximum limitation including the longitudinal force and steering angle command from the wheel.

In general, the ego vehicle tends to crash with high severity risk objects without crash mitigation algorithm. This action may cause more people dying on the road. We need the algorithm which can detect the crash severity and choose the path can mitigate the crash.

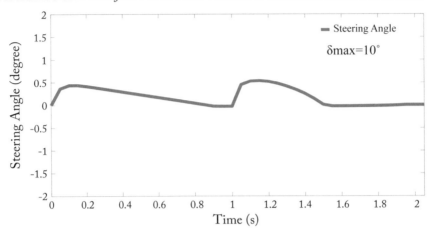

Figure 4.16: Steering angle command from the wheels of the ego vehicle before the crash.

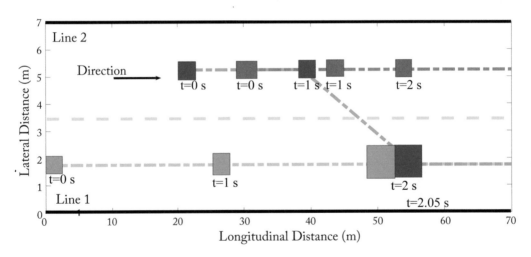

Figure 4.17: Positions of ego vehicle (green) approaching vehicle 1 (blue) and the following vehicle 2 (red).

(2) With Crash Mitigation

After implemented the mitigation crash algorithm, the ego vehicle selects to fully front crash with obstacle vehicle 1 with minimum crash severity, as shown in Fig. 4.17. Then there is no crash angle, which may cause less severity than crash has crash angle. The collision happens at 2 s, when ego vehicle is (51.00 m, 1.75 m) and the approaching vehicle 1 is (55.33 m, 1.75 m). The trajectory is marked every 1 s. The longitudinal velocity and force of the ego vehicle before the collision is shown in Fig. 4.18.

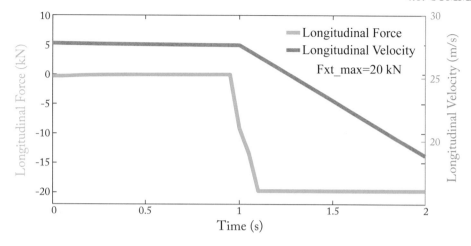

Figure 4.18: Longitudinal force and velocity of the ego vehicle before crash.

On account of the analysis above, we conclude that the ego vehicle can select the path with minimum crash severity when the collision is unavoidable. In this case, the ego vehicle selects fully front crash with the front vehicle instead of a crash with an angle which has higher severity of rollover and secondary crashes.

4.6 SUMMARY

A motion-planning method for autonomous vehicles confronting emergency situations where collision is unavoidable, generating a path to mitigate the crash severity as much as possible is proposed in this chapter. The assumption is made that the motion-planning module receives desired lane and velocity information from global-planning module; the vehicles states from the estimation module; the information of obstacles and the road boundaries from the perception module. In this case, the MPC algorithm apply to motion-planning. The presented crash severity factor and artificial potential field to describe the obstacles are integrated within the objective of the controller to avoid the generic obstacle and have the lowest crash severity, if the motion-planning system determines collision is inevitable. Furthermore, the vehicle dynamic is also considered as an optimal control problem.

Three scenarios were simulated to prove that the proposed MPC algorithm had the ability to avoid obstacles and, if the collision was inevitable, could mitigate the crash. Field testing of this proposed motion-planning method is in process and more urban situations should be analyzed in the future, such as emergency situations at the traffic lights.

CHAPTER 5

Design of Robust Feedback Controller for Path Tracking

5.1 MODEL-BASED CONTROL ALGORITHM FOR PATH TRACKING

The main purpose of path-tracking control for automated vehicles is to control the vehicle's lateral and longitudinal motion along the reference path which is regulated by the path planner. In past decades, a great deal of researches have been done on path-tracking algorithms, and some of them have been applied to automated vehicles in recent years. Commonly, the path-tracking algorithms can be classified in to two categories, the classical control algorithms already introduced in Chapter 3 and the model-based control that will be discussed in this chapter.

As the name implies, model-based controller utilizes the mathematic model to reflect and predict the vehicle motion. Thus, the vehicle model plays a crucial role for the controller design. For the path-tracking controller, the vehicle is generally regarded as a rigid body. When analyzing the vehicle dynamics characteristics, we pay particular attention on the longitudinal, lateral, and yaw motions [75]. Regarding the vehicle vertical motion, it mainly influences ride comfort, is always neglected in the path tracking control. Therefore, when establishing the path-tracking controller for automated vehicles, the plane motion model is widely adopted, which assumes vehicle moves on the flat road surface and the vertical suspension forces applied to the sprung mass are neglected or simplified as the load transfer caused by the inertia force. For different control requirements, geometric model, kinematic model, and dynamic model can be selected for usual control models. Based on the Ackermann steering configuration, the geometric model only concerns vehicle dimensions, which mainly reflects the relationship between vehicle position and road curvature during the turning maneuver. Considering the velocity and acceleration, the vehicle kinematic model not only describes the vehicle position, but also motion relationships without regarding the vehicle internal force [76]. On the contrary, for the vehicle dynamic model, the tire forces, mass inertia, and momentum of the system are considered by using Newton's second law. In addition to the geometric and kinematic relationships, the dynamic model can also illustrate the vehicle dynamic characteristics under different conditions, which is especially suitable for analyzing the vehicle stability during a path-tracking maneuver [77].

This chapter mainly discusses the path-tracking problem of autonomous vehicles in complex conditions and introduces the specific application of robust feedback control in path-tracking through a case study. Section 5.2 presents path-tracking control architecture, which includes the modeling of vehicle dynamic, system uncertainty analysis, and the design of the robust gain-scheduling lateral tracking controller. In Section 5.3, several cases are proposed to verify the tracking performance of the controller. Finally, Section 5.4 concludes the whole chapter and discusses future work.

5.2   AUTOMATED PATH TRACKING ARCHITECTURE

The control architecture for vehicle path-tracking is presented in this section. First, the modeling of vehicle is introduced. Then the basic theory of linear matrix inequality and the design of a robust path-tracking control algorithm is proposed.

5.2.1   MODELING

(1) Vehicle Dynamic Model

In reality, a vehicle model is a complex nonlinear coupling system with multi-degree of freedom. The accurate vehicle model is hard to be adopted for real-time control due to its high computation cost. With certain simplicity and assumptions, the single-track "bicycle" model is proposed in this chapter for the path-tracking controller design. The "bicycle" model has two speed states and three position states, and can adequately capture the tracking performance and handling stability under different conditions, shown in Fig. 5.1.

With the small angle assumption of the steering angle of front tire and the vehicle heading angle, the single-track vehicle plane dynamic is described as:

$$\begin{cases} \dot{Y} = V_x \psi + V_y \\ \dot{V}_y = \frac{1}{m} \left(F_{yf} + F_{yr} \right) - V_x r \\ \dot{\psi} = r \\ \dot{r} = \frac{1}{I_z} \left(l_f F_{yf} - l_r F_{yr} \right), \end{cases} \tag{5.1}$$

where m, I_z represent the vehicle mass and yaw inertia and l_f, l_r denote the distance from the vehicle CG to the front/rear axle, respectively. V_x and V_y are the vehicle longitudinal and lateral speed. F_{yf}, F_{yr} are the total lateral force of the front/rear tires, respectively. r is the vehicle yaw rate and ψ is vehicle heading angle.

(2) Tire Model

The tires are the main components of vehicles receiving external forces from the ground. The longitudinal and lateral tire forces directly affect the vehicle's acceleration/deceleration and stability performance. Therefore, the dynamic characteristics of the tire have a vital influence on the vehicle dynamic model. However, due to the complex relationship between the tire material and structure, the tire model has strong nonlinear characteristics. There have been many researches

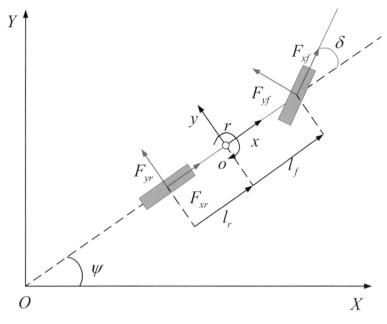

Figure 5.1: Bicycle vehicle model.

about the tire model, including brush model [78], "fiala" model [79], "uni-tire" model [80], magic formula tire model [81], and so on. Take the magic formula as the example; the tire lateral force can be expressed as:

$$y\left(\alpha\right) = D\sin\left(C\arctan\left\{B\alpha - E\left(B\alpha - \arctan\left(B\alpha\right)\right)\right\}\right), \tag{5.2}$$

where $y(\alpha)$ is the lateral tire force and α represents tire side-slip angle. B, C, D, and E are the stiffness factor, shape factor, peak factor, and curvature factor, respectively. The tire lateral force with different tire slip angle and tire load can be seen in Fig. 5.2.

However, to some extent the magic formula tire model is somewhat difficult for the controller design. When the tire slip angle is small, we usually assume that the tire-cornering stiffness keeps constant. Thus, the tire model can be simplified to the linear model as:

$$F_{yf} = C_f\alpha_f = C_f\left(\delta - \frac{V_y + l_f r}{V_x}\right) \tag{5.3}$$

$$F_{yr} = C_r\alpha_r = C_r\left(-\frac{V_y - l_r r}{V_x}\right), \tag{5.4}$$

where C_f, C_r are the cornering stiffness of the front and rear axle. α_f and α_r are the front and rear tire slip angle. δ is the front steering wheel angle.

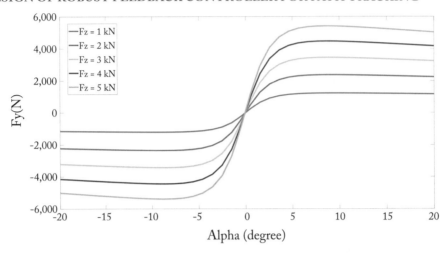

Figure 5.2: Nonlinear tire model.

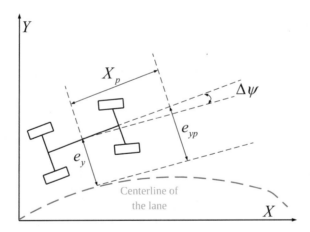

Figure 5.3: Single-point preview model for path tracking.

(3) Preview Tracking Model

In this chapter, the single-point preview model is applied for the path-tracking, which is shown in Fig. 5.3.

The single-point preview model is established based on the vehicle kinematic. In Fig. 5.3, e_y and $\Delta\psi$ denote the lateral error and heading angle error, X_p is the prediction distance, and the prediction lateral error e_{yp} is calculated by e_y and $\Delta\psi$. The kinematic relationship is expressed as:

$$\begin{cases} \dot{e}_y = V_x \sin \Delta\psi + V_y \cos \Delta\psi \\ e_{yp} = e_y + X_p \sin \Delta\psi \\ \Delta\dot{\psi} = \dot{\psi} - \dot{\psi}_d = r - k(s)\dot{s} \end{cases}, \tag{5.5}$$

where s denotes the curvilinear coordinate (arc-length) of point T along the path from the predetermined initial position. $k(s)$ represents the curvature of the reference path at the point T.

When the heading angle $\Delta\psi$ is small, it can be regarded as $\Delta\psi = \sin \Delta\psi$; in addition, compared with the longitudinal speed V_x, lateral speed V_y is small enough to be neglected. Utilizing the linear tire model, we can combine the vehicle dynamic model and preview tracking model. We define the system state as $x = [x_1 \ x_2 \ x_3 \ x_4]^T$ with $x_1 = e_y, x_2 = \Delta\psi, x_3 = V_y, x_4 = r$. We also define the steering wheel angle is the control input of th system as $u = \delta_f$ and $w = k(s)$ as the reference. Then the system can be described as:

$$\dot{x} = Ax + B_1 u + B_2 w, \tag{5.6}$$

where

$$A = \begin{bmatrix} 0 & V_x & 1 & 0 \\ 0 & 0 & 0 & 1 \\ 0 & 0 & -\dfrac{C_f + C_r}{mV_x} & -V_x - \dfrac{C_f l_f - C_r l_r}{mV_x} \\ 0 & 0 & \dfrac{C_r l_r - C_f l_f}{I_z V_x} & -\dfrac{C_f l_f^2 + C_r l_r^2}{I_z V_x} \end{bmatrix}, \quad B_1 = \begin{bmatrix} 0 \\ 0 \\ \dfrac{C_f}{m} \\ \dfrac{C_f l_f}{I_z} \end{bmatrix}, \quad B_2 = \begin{bmatrix} 0 \\ 1 \\ 0 \\ 0 \end{bmatrix}.$$

5.2.2   FEEDBACK CONTROL SCHEME

For the lateral path-tracking controller, its main purpose is to generate the corresponding steering angle input through state feedback to eliminate the tracking error when the road curvature changes. The structure of the feedback controller is shown in Fig. 5.4.

For the real systems, the error caused by model uncertainty and external disturbance is always inevitable. Then the goal of the robust feedback controller is to design a feedback gain matrix K to maintain the system have the relatively small output $z(t)$ against the external disturbances and model uncertainty. From Equation (5.6), it's obvious that the road curvature change is regarded as the external disturbance in the vehicle path-tracking control system. Moreover, in the state matrix A and B, the model uncertainty and parameter perturbation also exist. First, the tire-cornering stiffness is a typical uncertain factor. For simplicity, we regarded the tire force is linear with the tire slip angle. It can be seen from Fig. 5.3 that the linear tire model is appropriate in a certain range. However, we can't guarantee that the tire always stays in such linear range. When a sharp steer occurs, the tire loses its linear characteristics. The tire-cornering stiffness has the substantial reduction. If we still adopt the same cornering stiffness with the linear region,

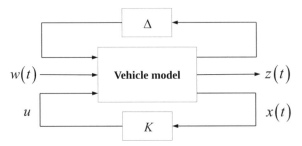

Figure 5.4: The architecture of the robust feedback controller.

the controller performance will deteriorate violently. Simultaneously, the different tire load and road adhesive coefficient also affect the tire cornering stiffness. Thus, the tire-cornering stiffness is regarded as the uncertain factor.

For the tire-cornering stiffness uncertainty, it can be expressed as:

$$C_f = C_{f0} + \lambda_f \Delta C_f \tag{5.7}$$

$$C_r = C_{r0} + \lambda_r \Delta C_r, \tag{5.8}$$

where λ_f, λ_r are the time-varying parameters satisfying $|\lambda_f| < 1$ and $|\lambda_r| < 1$. C_{f0}, C_{r0} represent the nominal front/rear tire cornering stiffness. So the system matrix can be written as:

$$A = A_0 + \Delta A \tag{5.9}$$

$$B_1 = B_{10} + \Delta B, \tag{5.10}$$

where A_0, B_{10} are the nominal system matrix. ΔA and ΔB_1, reflect the change of the matrix caused by cornering stiffness uncertainty.

Substitute Equations (5.9) and (5.10) into Equation (5.6), the system state space equation can be written as:

$$\dot{x} = (A_0 + \Delta A)x + (B_1 + \Delta B)u + B_2 w. \tag{5.11}$$

Here, we further define the ΔA and ΔB_1 as:

$$\Delta A = H \Lambda E_1 \tag{5.12}$$

$$\Delta B_1 = H \Lambda E_2, \tag{5.13}$$

where:

$$H = \begin{bmatrix} 0 & 0 \\ 0 & 0 \\ 1 & 0 \\ 0 & 1 \end{bmatrix}, \quad \Lambda = \mathrm{diag}\{\lambda_f, \lambda_r\}$$

$$E_1 = \begin{bmatrix} 0 & 0 & -\dfrac{(\Delta C_f + \Delta C_r)}{mv_x} & -\dfrac{(\Delta C_f l_f - \Delta C_r l_r)}{mv_x} \\ 0 & 0 & \dfrac{(-\Delta C_f l_f + \Delta C_r l_r)}{I_z v_x} & -\dfrac{(\Delta C_f l_f^2 + \Delta C_r l_r^2)}{I_z v_x} \end{bmatrix}, \quad E_2 = \begin{bmatrix} \dfrac{\Delta C_f}{m} \\ \dfrac{\Delta C_f l_f}{I_z} \end{bmatrix}.$$

Except for the uncertain factor, there have some other time-varying parameters in the system matrix. Such parameters should also be considered as the system uncertainty. During the path-tracking process, the vehicle longitudinal speed varies at any time. The vehicle should decelerate when obstacle occurs or road curvature changes and accelerate to the desired speed when the road is straight. Therefore, it is difficult to ensure a vehicle driving at a constant speed all the time. In this section, we don't discuss the vehicle speed regulation method and vehicle speed control algorithm. We only analyze the effect of varying speed on the lateral path-tracking controller. Note that in the state matrix A, the vehicle longitudinal speed Vx and its reciprocal $1/Vx$ are the time-varying parameters. For general conditions, the vehicle longitudinal speed always varies within a fixed range. So assuming the longitudinal speed is varying in the range of $[V_{x\,min}, V_{x\,max}]$, then $1/Vx$ varies in $[1/V_{x\,max}, 1/V_{x\,min}]$. By defining $\tilde{\vartheta}_1 = 1/V_{x\,max}$, $\tilde{\vartheta}_2 = 1/V_{x\,min}$, and replacing Vx and $1/Vx$ in matrix A by $\tilde{\vartheta}_1$ and $\tilde{\vartheta}_2$, we define:

$$A_i = \begin{bmatrix} 0 & \dfrac{1}{\tilde{\vartheta}_i} & 1 & 0 \\ 0 & 0 & 0 & 1 \\ 0 & 0 & -\dfrac{\tilde{\vartheta}_i\,(C_f + C_r)}{m} & -\dfrac{1}{\tilde{\vartheta}_i} - \dfrac{\tilde{\vartheta}_i\,(C_f l_f - C_r l_r)}{m} \\ 0 & 0 & \dfrac{\tilde{\vartheta}\,(C_r l_r - C_f l_f)}{I_z} & -\dfrac{\tilde{\vartheta}\,(C_f l_f^2 + C_r l_r^2)}{I_z} \end{bmatrix}, \quad i = 1, 2. \qquad (5.14)$$

For the time-varying parameter $1/Vx$, it can be written as:

$$\frac{1}{V_x} = \sum_{i=1}^{2} h_i(t)\,\tilde{\vartheta}_i, \qquad (5.15)$$

where

$$h_1 = \frac{1/V_{x\,min} - 1V_x}{1/V_{x\,min} - 1/V_{x\,max}}, \quad h_2 = \frac{1/Vx - 1/V_{x\,max}}{1/V_{x\,min} - 1/V_{x\,max}},$$

when the longitudinal speed varies between $V_{x\,min}$ and $V_{x\,max}$, always satisfies:

$$\sum_{i=1}^{2} h_i(t) = 1, \quad h_i(t) \geq 0. \qquad (5.16)$$

For the path-tracking control, the goal is to eliminate the lateral position error and heading angle error to improve the tracking accuracy. Simultaneously, to enhance the vehicle stability, the lateral speed V_y should be as small as possible. Hence, the control output is chosen as $z(t) = \left[e_{yp}, e_\psi, V_y\right]^T$. The robust H_∞ state-feedback controller is proposed for lateral path-tracking, which is defined as:

$$u(t) = Kx(t), \tag{5.17}$$

where K is the feedback gain matrix. The vehicle states including vehicle position, speed, and yaw rate could be acquired by GPS/IMU system and the road information is obtained by a high precision map. With such state parameters, the ideal steering wheel angle can be calculated.

Substituting Equations (5.11)–(5.17) into the state space equation, the feedback control scheme can be rewritten as:

$$\dot{x}(t) = \sum_{i=1}^{2} h_i(t)\left(\bar{A}_i + H\Lambda\bar{E}_i\right)x(t) + B_2w(t)$$
$$z(t) = Cx(t), \tag{5.18}$$

where

$$\bar{A}_i = A_{0i} + B_{10}K_i, \quad \bar{E}_i = E_{1i} + E_2K_i$$

$$C = \begin{bmatrix} 1 & x_p & 0 & 0 \\ 0 & 1 & 0 & 0 \\ 0 & 0 & 1 & 0 \end{bmatrix}.$$

5.2.3   ROBUST CONTROLLER DESIGN

The control objective is to generate the control signal to maintain the system is stable and has H_∞ disturbance attenuation performance in the presence of external disturbance and system uncertainties:

$$\int_0^t z^T(t)z(t)\,dt \leq \gamma^2 \int_0^t w^T(t)w(t)\,dt, \tag{5.19}$$

where γ is the prescribed attenuation level.

In order to deal with the external disturbance and uncertain factors, we first introduce the following lemmas.

Lemma 5.1 [82] *For a time-varying diagonal matrix $\Phi(t) = \mathrm{diag}\left\{\sigma_1(t), \sigma_2(t), \ldots, \sigma_p(t)\right\}$ and two matrices R and S with appropriate dimensions, if $|\Phi(t)| \leq V$, where $V > 0$ is a known diagonal matrix. Then for any scalar $\varepsilon > 0$, we have:*

$$R\Phi S + S^T\Phi^T R^T \leq \varepsilon RVR^T + \varepsilon^{-1}S^T VS. \tag{5.20}$$

Lemma 5.2 **[83]**   *For given symmetric matrix* $S = \begin{bmatrix} S_{11} & S_{12} \\ S_{21} & S_{22} \end{bmatrix}$, *where the dimension of the* S_{11}
is $r \times r$, *the following conditions are equivalent:*

$$S < 0$$
$$S_{11} < 0, \quad S_{22} - S_{21} S_{11}^{-1} S_{12} < 0 \qquad (5.21)$$
$$S_{22} < 0, \quad S_{11} - S_{12} S_{22}^{-1} S_{12}^{T} < 0.$$

Theorem 5.3   *Given the positive constant positive scalars* γ, ρ, *the closed-loop system in Equation* (5.20) *is asymptotically stable with* $w(t) = 0$, *and satisfies the* H_{∞} *performance for all* $w(t) \in [0, \infty)$, *if there exists a symmetric positive definite matrix* X, *a positive definite matrix* W_i, *and positive scalars* ε_{ai} *such that:*

$$\begin{bmatrix} X^T A_i^T + W_i^T \bar{B}_1^T + A_i^T X + \bar{B}_1 W_i & B_2 & X^T C^T & \varepsilon_i H & X^T \bar{E}^T \\ * & -\gamma^2 I & 0 & 0 & 0 \\ * & * & -I & 0 & 0 \\ * & * & * & -\varepsilon_{ai} I & 0 \\ * & * & * & * & -\varepsilon_{ai} I \end{bmatrix} \leq 0, \quad i = 1, 2 \qquad (5.22)$$

$$\begin{bmatrix} -I & \sqrt{\rho} W_i \\ * & -u_{\max}^2 X \end{bmatrix} \leq 0, \quad i = 1, 2, \qquad (5.23)$$

where $u_{\max}$ *is the maximum steering wheel angle and* $*$ *represents the symmetric elements in the symmetric matrix.*

Now the derivation and proof process are given as follows.

Proof. Define the Lyapunov function as:

$$V(t) = x^T(t) P x(t), \qquad (5.24)$$

then

$$\dot{V} = \dot{x}^T P x + x^T P \dot{x}$$
$$= \sum_{i=1}^{2} h_i(t) \left(\left(\bar{A}_i + H \Lambda \bar{E}_i \right) x + B_2 w \right)^T P x + x^T P \left(\left(\bar{A}_i + H \Lambda \bar{E}_i \right) x + B_2 w \right). \qquad (5.25)$$

By adding $z^T z - \gamma^2 w^T w$ on both sides of the equation, we can get:

$$
\begin{aligned}
& \dot{V} + z^T z - \gamma^2 w^T w \\
& = \sum_{i=1}^{2} h_i(t) \left(\left(\bar{A}_i + H\Lambda\bar{E}_i \right) x + B_2 w \right)^T Px + x^T P \left(\left(\bar{A}_i + H\Lambda\bar{E}_i \right) x + B_2 w \right) \\
& \quad + z^T z - \gamma^2 w^T w \\
& = \sum_{i=1}^{2} h_i(t) \left(x^T \left(\Theta_i + C^T C \right) x + w^T B_2 Px + x^T PB_2 w - \gamma^2 w^T w \right) \\
& = \sum_{i=1}^{2} h_i(t) \begin{bmatrix} x \\ w \end{bmatrix}^T \begin{bmatrix} \Theta_i + C^T C & PB_2 \\ B_2^T P & -\gamma^2 I \end{bmatrix} [\, x \quad w \,],
\end{aligned}
\tag{5.26}
$$

where

$$
\Theta_i = P\bar{A}_i + PH\Lambda\bar{E}_i + \left(P\bar{A}_i + PH\Lambda\bar{E}_i \right)^T .
$$

It is obvious that the system is asymptotically stable and the H_∞ performance is satisfied only the following inequality holds:

$$
\sum_{i=1}^{2} \begin{bmatrix} \Theta_i + C^T C & PB_2 \\ B_2^T P & -\gamma^2 I \end{bmatrix} < 0.
\tag{5.27}
$$

By adopting the Schur complement, Equation (5.27) is equivalent to:

$$
\begin{bmatrix} \Theta_i & PB_2 & C^T \\ * & -\gamma^2 I & 0 \\ * & 0 & -I \end{bmatrix} \leq 0 \quad i = 1, 2.
\tag{5.28}
$$

Here, we define that $sys\{\bullet\}$ represents $\bullet + \bullet^T$, then above inequality can be deformed as:

$$
\begin{aligned}
& \begin{bmatrix} \Theta_i & PB_2 & C^T \\ * & -\gamma^2 I & 0 \\ * & 0 & -I \end{bmatrix} \\
& = \begin{bmatrix} sys\{P\bar{A}_i\} & PB_2 & C^T \\ * & -\gamma^2 I & 0 \\ * & 0 & -I \end{bmatrix} + sys\left\{ \begin{bmatrix} P\bar{H} \\ 0 \\ 0 \end{bmatrix} \Lambda [\, \bar{E}_i \quad 0 \quad 0 \,] \right\}.
\end{aligned}
\tag{5.29}
$$

Since $|\lambda| \leq 1$, $\Lambda\Lambda^T \leq I$, it follows Lemma 5.1 and Lemma 5.2 that the following inequality holds:

$$
\begin{bmatrix}
(A_{0i} + B_{01}K_i)^T P + P(A_{0i} + B_{01}K_i) & PB_2 & C^T & \varepsilon P\bar{H} & \bar{E}_i^T \\
* & -\Upsilon^2 I & 0 & 0 & 0 \\
* & * & -I & 0 & 0 \\
* & * & * & -\varepsilon I & 0 \\
* & * & * & * & -\varepsilon I
\end{bmatrix} \leq 0. \quad (5.30)
$$

Note that there are nonlinear terms in (5.30); the matrix P and K are coupled. It is hard to solve this inequality by linear matrix inequation. Here we define $P^{-1} = X$ and $K_i = W_i X^{-1}$, and perform a congruence transformation with $\Gamma = \text{diag}(X, I, I, I, I)$ to (5.30), then the inequality (5.30) is equivalent to the inequality (5.22) in Theorem 5.3.

Moreover, the control input is limited by the maximum steering wheel angle, it is possible to obtain $x^T(t) P x(t) < \rho$, and consider that:

$$
\begin{aligned}
\max |u(t)|^2 &= \max \left\| x^T(t) K_i^T K_i x(t) \right\| \\
&= \max \left\| x^T(t) P^{\frac{1}{2}} P^{-\frac{1}{2}} K_i^T K_i P^{-\frac{1}{2}} P^{\frac{1}{2}} x(t) \right\| \qquad (5.31) \\
&< \rho \cdot \theta_{\max} \left(P^{-\frac{1}{2}} K_i^T K_i P^{-\frac{1}{2}} \right) \leq u_{\max}^2 I,
\end{aligned}
$$

where $\theta_{\max}$ denotes the max eigenvalue. It can be deduced that (5.32) holds only if:

$$
\rho \left(P^{-\frac{1}{2}} K_i^T K_i P^{-\frac{1}{2}} \right) \leq u_{\max}^2 I \qquad (5.32)
$$

which can be written as

$$
\begin{bmatrix}
-I & \sqrt{\rho} W_i \\
* & -u_{\max}^2 X
\end{bmatrix} \leq 0. \qquad (5.33)
$$

Equation (5.33) is equivalent to Equation (5.23) in Theorem 5.3, and then the proof is completed. $\qquad\square$

The final feedback gain matrix can be written as:

$$
K = \sum_{i=1}^{2} h(i) K_i, \qquad (5.34)
$$

where K_i has been calculated off-line. The final feedback gain matrix changes only dependent on $h_1(t)$ and $h_2(t)$, which can be computed online with longitudinal speed V_x. It indicates that the proposed control algorithm is easy for real-time application.

Table 5.1: Parameters of autonomous vehicle and the controller

Parameter	Value	Unit	Parameter	Value	Unit
m	2270	kg	δ_{max}	10	deg
h	0.647	m	C_f	127,000	N
l_f	1.421	m	C_r	130,000	N
l_r	1.434	m	V_{max}	120	km/h
I_z	4600	kgm^2	V_{min}	0	km/h

5.3   CASE STUDY

To verify the feasibility and tracking performance of proposed robust feedback controller, three scenarios are carried out for simulations; the detailed vehicle parameters are shown in Table 5.1.

Scenario 1

This scenario is set up for the vehicle drives along the ramp into the highway. The road adhesion coefficient is set as 0.6 and the parameter uncertainty on the tire cornering stiffness is set as 40% of the nominal value. The initial longitudinal speed is 60 km/h. Note that the longitudinal control algorithm is designed as [84]. The vehicle adjusts the speed according to the road adhesion coefficient and curvature to guarantee the lateral stability. The simulation results are shown in Fig. 5.5.

It can be seen from Fig. 5.5, when the road curvature increases, the brake cylinders begin to generate the braking pressure to decelerate the vehicle in order to guarantee the vehicle stability. During the whole simulation process, the vehicle speed is always restricted under the safety threshold which effectively prevents the vehicle side slip. For the lateral controller, even the vehicle speed varies between about 33–60 km/h and the tire-cornering stiffness has 40% uncertainty, the steering wheel angle is also smooth and stable for the whole time without overshooting. The lateral error is limited less than 1 m in this scenario, indicating that the vehicle can follow the reference path accurately.

Scenario 2

This scenario is a "U-turn" maneuver on the low adhesion road which simulates that driving on the slippery road with large curvature. In this case, the initial vehicle speed is 50 km/h and the road adhesion coefficient is 0.3. The parameter uncertainty on the tire cornering stiffness is also set as 40% of the nominal value as scenario 1. In order to further reflect the robustness to the system uncertainty of the proposed robust feedback controller, another $H\infty$ feedback controller

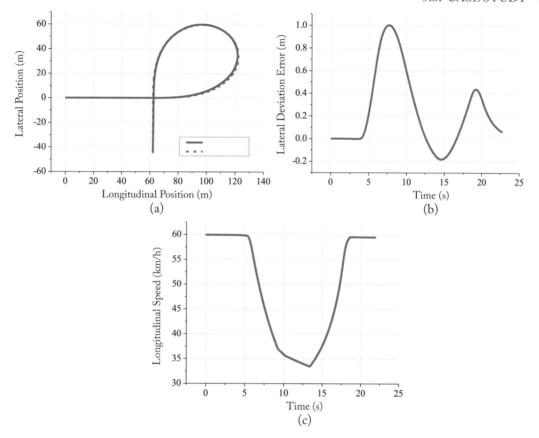

Figure 5.5: The simulation results for scenario 1.

without considering the tire stiffness uncertainty is introduced here, which is presented by the following corollary.

Corollary 5.4 *Given the positive constant positive scalars γ, ρ, the closed-loop system is asymptotically stable with $w(t) = 0$, and satisfies the H_∞ performance for all $w(t) \in [0, \infty)$, if there exists a symmetric positive definite matrix X, a positive definite matrix W_i such that:*

$$\begin{bmatrix} X^T A_i^T + W_i^T \bar{B}_1^T + A_i^T X + \bar{B}_1 W_i & B_2 & X^T C^T \\ * & -\gamma^2 I & 0 \\ * & * & -I \end{bmatrix} \leq 0 \ \ i = 1, 2 \qquad (5.35)$$

$$\begin{bmatrix} -I & \sqrt{\rho} W_i \\ * & -u_{\max}^2 X \end{bmatrix} \leq 0 \ \ i = 1, 2. \qquad (5.36)$$

Table 5.2: Comparison of controller A and controller B

	Speed Variation	Tire stiffness uncertainty
Controller A	✓	✓
Controller B	✓	✗

Then the feedback control gain matrix is given by:

$$K = \sum_{i=1}^{2} h(i) W_i X. \tag{5.37}$$

The proof process of Corollary 5.4 is not described here, which is similar with the proof for Theorem 5.3. Now we define the proposed H_∞ robustness feedback controller as the controller A and the H_∞ feedback controller as the controller B.

Figure 5.6a–f illustrate the control performance of each controller for scenario 2. It can be seen that for controller A and controller B, they both begin to decelerate when the road curvature increases and exceeds the safety threshold. The vehicle is effectively prevented from side slip through these two controllers, and the tracking performances are all acceptable. However, the lateral controller for controller B is designed under the condition that the tire stiffness is accurate. When the parameter uncertainty exists, the calculated steering angle can't generate accurate lateral tire force as the demand signal. When the cornering stiffness is 40% of nominal value, the vehicle with controller B deviates from the reference path and its deviation error is greater than that of controller A. From Fig. 5.6c, it can be seen that the maximum lateral deviation for controller B reaches 1.97 m, while that of controller A is only 0.33 m. Therefore, for scenario 2, the controller A has better tracking performance under the parameter varying and system uncertainty condition.

Scenario 3

This scenario is following a slalom trajectory on high adhesion road which simulates the continue obstacles avoidance. In this scenario, the initial vehicle speed is 80 km/h and the road adhesion coefficient is 0.85. The parameter uncertainty on the tire cornering stiffness is set as 70% of the nominal value.

Figure 5.7 shows the simulation results of each controller for scenario 3. Note that the adhesion coefficient is 0.85 for this case, and the initial speed doesn't exceed the safety threshold during the whole process. Figure 5.7a shows the tracking performance for these two controllers of different system uncertainty status (no uncertainty and 70% uncertainty). The reference path is like a "sin-wave" that requires vehicle continues switching the direction of the steering wheel. When the cornering stiffness with no uncertainty (the actual cornering stiffness equals to the

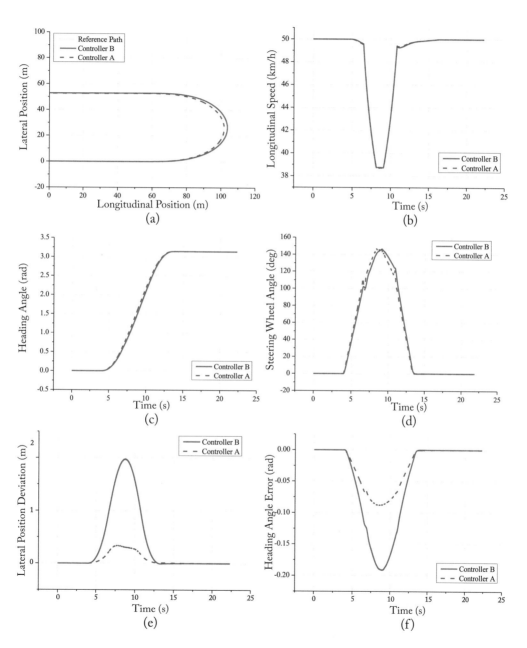

Figure 5.6: The simulation results for scenario 2.

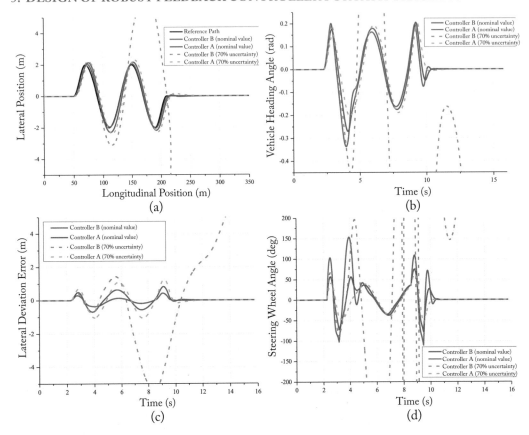

Figure 5.7: The simulation results for scenario 3.

nominal value), both controller A and controller B have good tracking performance. Even controller B has higher tracking accuracy, the maximum lateral error is only 0.33 m, which is 0.63 m for controller A. When the severe uncertainty exists (the actual cornering stiffness is only 70% of the nominal value), the tracking performances for both controller A and controller B are deteriorated. The maximum lateral error for controller A reaches 1.1 m. However, it still can follow the reference path. For controller B, the attenuation of actuator response causes the increase of tracking error, which leads to the overshoot of control signal at 4.7 s, and leads to vehicle lose its stability finally. As mentioned before, controller B is designed by Corollary 5.4, which implies that it could have best control performance when the actuator is healthy but can't guarantee system stability with disturbance or actuator fault happens. On the other hand, controller A is calculated by solving linear matrix inequation with a larger dimension than controller B. Thus, it sacrifices some tracking accuracy for a wider stable region of the system.

5.4 SUMMARY

Due to the strong nonlinear and coupling characteristics, the vehicle model should be linearized and simplified for the controller design. However, some parameters of the vehicle are time-varying and difficult to identified, which will bring uncertainty to the linear vehicle model and affect the control accuracy. For some severe conditions, model uncertainty may cause control overshooting and vehicle instability. In this chapter, a feedback gain-scheduling controller is proposed based on the $H\infty$ control theory. The linear matrix inequality is established considering the varying parameter (longitudinal speed), system uncertainty (tire cornering stiffness), and output constraint, and the modeling error is regarded as the disturbances. The tracking performance of the proposed controller is evaluated via a co-simulation platform established by using MATLAB and CarSim software. The simulation results indicate that the controller proposed in this chapter can slow down the vehicle in time and maintain the lateral stability of the vehicle under the condition of large curvature and low adhesion coefficient, and has satisfactory tracking accuracy even system uncertainty exists.

CHAPTER 6

Collision Avoidance in Longitudinal Direction With/Without V2X Communication

6.1 INTRODUCTION

There were nearly 2 million police-reported front-to-rear crashes in 2017, representing 32% of all crashes [85]. Front crash prevention systems, which warn drivers, brake autonomously, or perform both functions when a frontal collision is imminent, have been estimated to potentially prevent or mitigate up to 70% of front-to-rear collisions and 20% of all police-reported crashes if installed on all passenger vehicles [86]. In addition, in some special cases, such as when a moose suddenly crosses the road in front of the vehicle, front-crash prevention systems can reduce the severity of a crash by lowering the speed of the host vehicle, as shown in Fig. 6.1.

A few types of front-crash prevention systems are available to consumers. Forward Collision Warning (FCW) was first introduced in the United States by Mercedes-Benz in 2000. Systems with both FCW and Autonomous Emergency Braking (AEB) followed, and were first offered in the United States by Acura in 2006. Most systems were initially offered as optional equipment in luxury vehicles but have become more widely available in recent years. Volvo offered an AEB system that operates only at low speeds as standard equipment on the XC60 in 2010. In 2016, 54% of vehicle series in the United States offered a front-crash prevention system as optional equipment and 7% included one as standard. Twenty automakers representing 99% of the auto market in the United States have committed to making FCW and AEB standard features on virtually all new passenger vehicles by 2022.

In recent years, many researches have studied the perception and control technologies of FCW and AEB systems to improve traffic safety. However, most of the previous studies on collision avoidance and road safety only considered the impact of the nearest two vehicles on the road [87, 88]. This is because the input signals for collision avoidance systems are mainly based on on-board sensors (e.g., radar, camera or LiDAR), without inter-vehicle communications. The shortcoming of such systems is that the collective behavior of multiple closely spaced vehicles (i.e., a coupled group) in string formation may cause vehicular accidents on highways,

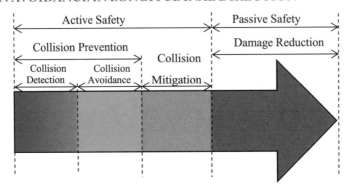

Figure 6.1: Front crash prevention and mitigation.

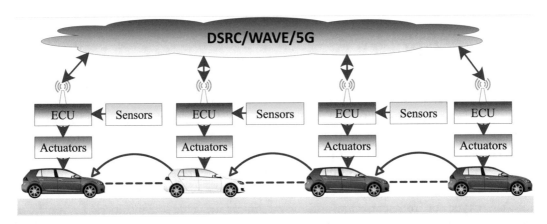

Figure 6.2: Overall configuration of multiple-vehicle collision avoidance.

particularly under congested conditions, because the last vehicle in the formation experiences an accumulated time delay in responding to the behavior of the vehicles in the front.

In fact, a string of coupled vehicles on a highway is more likely to result in multiple-vehicle accidents, which is considered difficult for conventional longitudinal collision avoidance systems to solve. With the rapid development of wireless communication and distributed computing technology, the Cooperative Collision Avoidance Systems (CCAS) provides a possible way to further address this severe issue. The CCAS has a great potential for avoiding such serious accidents or, at least, for minimizing the impact if the collision is unavoidable by simultaneously controlling the braking of multiple vehicles. In V2V systems, the use of Dedicated Short Range Communication (DSRC), Wireless Access in Vehicular Environments (WAVE), or the fifth generation of cellular network technology (5G) has the overall advantages of extensive network radio communication capabilities that demonstrate low-latency and high throughput, and which is both robust and scalable, as shown in Fig. 6.2. Due to these kinds of characteristics,

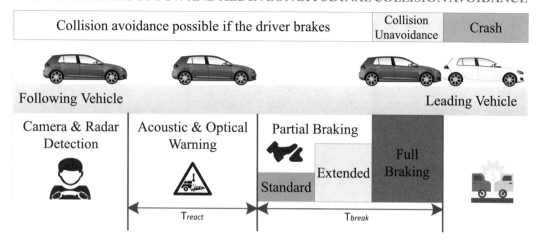

Figure 6.3: Steps and system components associated with the forward collision avoidance.

DSRC/WAVE/5G is suitable for improving the road traffic safety under the vehicle networking environment, and can be also combined with other on-board sensor systems, such as environmental perception systems (e.g., LiDAR/radar), precision position systems (e.g., GPS/DGPS), and vehicle state estimation systems (e.g., wheel speed, vehicle velocity, acceleration and yaw rate). In addition, the relative position and speed of vehicles can be obtained and shared by multisource information fusion or communication, which makes the application of CCAS and Intelligent Connected Vehicles (ICVs) in collision avoidance possible.

In this chapter, we will introduce the basic principles and methods of FCW and AEB systems as well as their specific applications in prevention of multi-vehicle pile-up crashes based on V2X communication.

6.2   EFFECTIVENESS OF FCW AND AEB IN LONGITUDINAL COLLISION AVOIDANCE

Longitudinal collision avoidance technologies attempt to prevent front-to-rear accidents, which are usually due to the difficulty for the driver to react to a sudden braking by the leading vehicle. A longitudinal collision avoidance system operates in the following manner: different types of sensors, such as cameras, radars, or LiDAR, are installed at the front of a vehicle to continuously detect the potential conflicts on road, such as a slow-moving or stopped vehicle. When one is found, the system determines whether the vehicle is in imminent danger of crashing, and if so, it begins the process of alerting the driver through different warning cues and initially preparing the active brakes control system. If the conflict persists, the system provides additional braking force if the driver brakes too late or not strongly enough [89], as shown in Fig. 6.3.

The typical criteria for activation of longitudinal collision avoidance system are as follows:

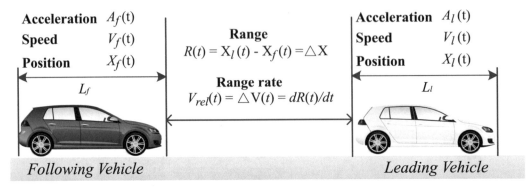

Figure 6.4: Variables in the vehicle-following situation.

(a) Worst-case scenario: the system assumes that the leading vehicle could stop at any time. In essence, it maintains a "safe distance, $R(t)$," i.e., the minimum distance necessary to come to a stop in the event the leading vehicle suddenly stops.

(b) Time-to-collision (TTC): the system determines the time required for two vehicles to collide if they continue at their current speed and on the same path (assuming that V_f is greater than V_l).

Figure 6.4 shows a vehicle-following situation and the considered variables.

As reported by the literature, several algorithms have been developed to activate the collision avoidance systems and in particular the two main types are based on kinematic and perceptual approach. Whatever the criterion of activation, the system must be reliable, functional, and secure, which can also be shared by the driver (i.e., it should give few false alarms). In the following sections we analyze these two different approaches, which are based on different assumptions and therefore operate in different ways.

6.2.1   KINEMATIC APPROACH

The algorithms based on kinematic trigger alerts use the fundamental laws of motion. Combining the hypothesis of the deceleration and reaction time with the current state of a vehicle, the algorithm determines a minimum distance required to stop safely. When the vehicle is at a distance minor or equal to the safe distance from another vehicle, an alarm is triggered. By and large, the kinematic approach starts the warning based on the criterion of the worst case.

Different algorithms were adopted to calculate a critical warning distance based on vehicle motion and the variables related to human characteristics (vehicle speed, acceleration, delay in the human reaction, etc.). When the distance is smaller than the limit value calculated according to the type of system used, an alarm is activated. Table 6.1 shows the equations and terms most frequently used in collision avoidance systems.

Table 6.1: Frequently used equations and terms in collision avoidance systems

Meaning	Form
Following vehicle stopping distance (R_f)	$R_f = -\dfrac{V_f^2}{2A_f}, A_f < 0, V_f \geq 0$
Leading vehicle stopping distance (R_l)	$R_l = -\dfrac{V_l^2}{2A_l}, A_l < 0, V_l \geq 0$
Reaction time margin (R_{td})	$R_{td} = \tau V_f, \quad \tau > 0, V_f > 0$
Range rate margin (R_{rr})	$R_{rr} = \tau V_{rel}, \quad \tau > 0, V_{rel} > 0$
Minimum range (R_{min})	$R_{min} = \text{constant} > 0$

In Table 6.1, $V_{rel} = \Delta V$ is the relative speed between the following and leading vehicles in m/s; for the other parameters see Fig. 6.4.

The main types of kinematic algorithms found in the literature are described below.

(1) The Mazda Algorithm

The algorithm developed by Mazda considers a hypothetical worst case. The scenario assumes that initially the two vehicles maintain constant speeds: V_l and V_f. Subsequently, the leading vehicle starts to brake after time τ_2 with a deceleration rate A_l, while the host vehicle starts to brake after an additional time τ_1 at deceleration rate A_f, which continues until both vehicles come to a full stop. The algorithm shown below calculates the minimum space required to ensure that the scenario described above occurs without collisions.

$$R_{warn} = f\left(V_l, V_f, V_{rel}\right) = \frac{V_f^2}{2A_f} - \frac{V_l^2}{2A_l} + V_f \tau_1 + V_{rel}\tau_2 + R_{min}, \tag{6.1}$$

where R_{min} is the minimum range (see Fig. 6.4 and Table 6.1). The distance calculated by this equation must be taken as a critical warning distance. The values of the variables are reported in Table 6.2.

The danger distance of the Mazda algorithm is shown in Fig. 6.5. An implementation of the Mazda algorithm in MATLAB is given in Listing 6.1.

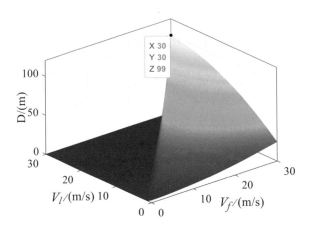

Figure 6.5: The Mazda algorithm.

Listing 6.1: Mazda algorithm

```
v=0:1:30;  vr=0:1:30;
a1=6; a2=8;%
t1=0.1; t2=0.6;
d0=3;
% Equation: dbr=0.5*(v^2/a1-(v-vr)^2/a2)+v*t1+vr*t2+d0;
for i=1:31
   for j=1:31
     if v(j)<vr(i)
        y(i,j)=0;
     else
        y(i,j)=0.5*((v(j)^2/a1)-(v(j)-vr(i))^2/a2)+v(j)*t1+vr(i)*t2+d0;
     end
   end
end
surf(v,vr,y); box on
shading interp;
title('Mazda Algorithm')
xlabel('v/(m·s-1)')
ylabel('vr/(m·s-1)')
zlabel('db/m')
```

Table 6.2: Parameter values for algorithms

	A_f (m/s^2)	A_l (m/s^2)	τ_1 (s)	τ_2 (s)	τ_d (s)	τ(s)	$R_{\min}$ (m)
Mazda	6.0	8.0	0.1	0.6	–	–	5.0
SDA	7.0	7.0	–	–	1.0	–	–
Berkeley	6.0	6.0	–	–	–	1.2	5.0

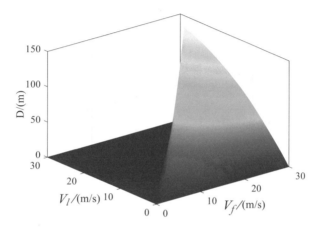

Figure 6.6: The stop distance algorithm.

(2) The Stop Distance Algorithm (SDA)
This algorithm defines a warning distance based on the difference between the stopping distances of the leading and following vehicles. If the distance between the two vehicles is less than the warning distance, an auditory collision warning alarm is presented to the driver. The following equation describes the logic activating the alarm:

$$R_{warn} = f\left(V_l, V_f\right) = V_f \tau_d + \frac{V_f^2}{2A_f} - \frac{V_l^2}{2A_l}, \tag{6.2}$$

where τ_d is the driver's reaction time. The values of the variables are reported in Table 6.2.

The danger distance of the stop distance algorithm is shown in Fig. 6.6. An implementation of the stop distance algorithm in MATLAB is given in Listing 6.2.

(3) The PATH Berkeley Algorithm
This algorithm is a modified version of the Mazda algorithm. The critical warning distance is given by:

$$R_{warn} = f\left(V_l, V_f\right) = \frac{V_f^2 - V_l^2}{2A_{\max}} + V_f \tau + R_{warn}. \tag{6.3}$$

Listing 6.2: The stop distance algorithm

```
v=0:1:30;  vr=0:1:30;
t=1; a=7;
% Equation: dbr=v*t+(v^2-(v-vr)^2)/(2a)
for i=1:31
    for j=1:31
    if v(j)>vr(i)
        y(i,j)=v(i)*t+(v(j)^2-(v(j)-vr(i))^2)/(2*a);
    else
        y(i,j)=0;
    end
    end
end
surf(v,vr,y); box on
title('Stop Distance Algorithm')
xlabel('v/(m·s-1)')
ylabel('vr/(m·s-1)')
zlabel('dbr/m')
```

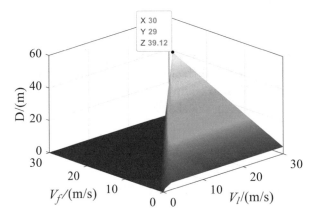

Figure 6.7: The PATH Berkeley algorithm.

According to this algorithm, the leading vehicle brakes at the maximum deceleration rate A_{max}, while the following vehicle starts to brake after reaction time τ at the same deceleration. R_{warn} is the minimum distance required to avoid collisions until both vehicles come to a full stop. The values of the variables are presented in Table 6.2.

Listing 6.3: The PATH Berkeley algorithm

```
v=0:1:30; vr=0:1:30;
t1=1; t2=0.2; a2=6;
%Equation: dbr=vr*(t1+t2)+0.5*a2*(t1+t2)^2
for i=1:31
   for j=1:31
   if v(j)>vr(i)
      y(i,j)=vr(i)*(t1+t2)+0.5*a2*(t1+t2)^2;
   else
      y(i,j)=0;
   end
   end
end
surf(v,vr,y); box on
title('Berkeley Algorithm)
xlabel('v/(m·s-1)')
ylabel('vr/(m·s-1)')
zlabel('dbr/m')
```

The danger distance of the PATH Berkeley algorithm is shown in Fig. 6.7. An implementation of the PATH Berkeley algorithm in MATLAB is given in Listing 6.3.

(4) CAMP FCW Algorithm

For the forward-collision warning system, the driver will be too nervous if he/she is alerted too often. Drivers were asked to execute last-second braking and steering maneuvers while approaching a target-leading vehicle. Based on the data recorded on field, the CAMP FCW algorithm was defined. It determines the alert range necessary to assist the driver to avoid a potential crash, which is a function of speeds and decelerations of the leading and following vehicles, and total delay time. The last parameter (total delay time) was the composite sum of three separate delay times: the interface delay time, the driver brake delay time and the brake system delay time. The interface delay time is defined as the time between when the crash alert criterion was violated and when the crash alert was presented to the driver. This delay is assumed to be 0.18 s. The driver brake delay is defined as the time between crash alert onset and when the driver triggered the brake switch. This delay was assumed to be 1.50 s for surprise alerts. The brake system delay time is defined as the time between braking onset and vehicle slowing and is assumed to be 0.20 s. The total delay time is approximately equal to 1.88 s:

$$R_{warn} = f\left(V_l, V_f, A_l, A_f, T_t,\right)$$
$$= \left[\left(V_f - V_l\right) T_t + \frac{\left(A_f - A_l\right) T_t^2}{2}\right]$$
$$- \frac{0.5\left[V_f - V_l + \left(A_f - A_l\right) T_t\right]^2}{\left[-1.61 + 0.668 A_l - 0.807\left(V_f - V_l\right) + 0.765\right] - A_l}. \qquad (6.4)$$

6.2.2   PERCEPTUAL APPROACH

The perceptual approach triggers an alarm based on the thresholds of perception, that is, when the human perceptual threshold is exceeded, a signal is activated to alert the driver. The Time to Collision (TTC) is considered as one of the most widely used safety time thresholds and a measure of a crash risk. It is defined as the time until a collision between two vehicles would have occurred if the collision course and speed difference were maintained:

$$TTC = \frac{X_f - X_l - L_f}{V_f - V_l}, \qquad (6.5)$$

where L_f is the length of the following vehicles; for the other variables see Fig. 6.4.

The TTC is only calculated if the following vehicle's speed is greater than the one of the leading vehicle ($V_f > V_l$). In particular, low TTC indicates a higher risk of collision. Usually, thresholds value for the TTC are set to trigger a warning, also called a critical value. The threshold value in the literature ranges between 2 and 5 s. The main types of perceptual algorithms found in the literature are described below.

(1) The Honda Algorithm

According to the Honda algorithm, a warning is triggered when the TTC is equal to 2.2 s. The distance is calculated by means of:

$$R_{warn} = f\left(V_{rel}\right) = 2.2 V_{rel} + 6.2, \qquad (6.6)$$

where 6.2 m is an additional safety value.

The danger distance of the Honda algorithm is shown in Fig. 6.8. An implementation of the Honda algorithm in MATLAB is given in Listing 6.4.

(2) The Honda's Collision Mitigation Braking System (CMBS)

Another system by Honda, called Collision Mitigation Braking System (CMBS), is also based on the evaluation of the TTC. This system returns three different alerts. There is an initial phase of attention if the distance between the following and leading vehicles exceeds the safety limit, estimated with a threshold of TTC equal to 3 s. When this threshold is exceeded, an alarm

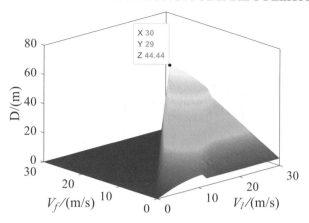

Figure 6.8: The Honda algorithm.

starts to suggest the driver to take preventive measures to avoid collision. After this, there are other phases of prevention and action that are activated if the distance continues to gradually decline, i.e., TTC less than 2 s and less than 1 s, respectively. In both cases the aim is to reduce the severity of impact, but the collision will hardly be avoided.

(3) The Hirst and Graham Algorithm
Many criticisms have been reported in the literature on the validity of the assumption of a constant threshold value of the TTC to distinguish between safe and unsafe situations. It is often suggested to investigate if and how this parameter varies depending on speed. In relation to the possible link between the TTC and speed, it is worth mentioning a perceptual-based algorithm developed by Hirst and Graham. They suggest that a TTC-based warning should be increased with a speed penalty (SP) to achieve appropriately timed warning. They suggest that this type of algorithm would minimize nuisance alarms. In essence, the algorithm developed uses a threshold set at 3 s of the TTC, with an adjustment for the vehicle speed (SP) equal to 0.4905 m for each kilometer per hour of the following vehicle speed:

$$R_{warn} = f\left(V_{rel}, V_f\right) = 3\frac{dR}{dt} + 0.4905V_f, \tag{6.7}$$

where $\frac{dR}{dt} = \Delta v$ is the relative speed between two vehicles.

6.2.3   AEB TEST SCENARIOS, PROCEDURES, AND SIMULATIONS
Analysis of real-world crash events enables the AEB system can be adapted to prevent the most common collisions on road. Reasonable test procedures design can simulate common scenarios of crash accidents and effectively test various performances of AEB system. The development of these procedures and simulations is summarized here. In China, the C-NCAP has developed

Listing 6.4: The Honda algorithm

```
v=0:1:30; vr=0:1:30;
t1=0.5; t2=1.5;
a1=7.8; a2=7.8;
% Equation If v2/a2>=t2 db1=t2*vr+t1*t2*a1-0.5*a1*t1^2;
%              If v2/a2<t2 db2=t2*v(i)-0.5*(t2-t1)^2-v2^2/(2*a2);
for i=1:31
   for j=1:31
      v2(i,j)=v(j)-vr(i);
   if v2(i,j)/a2>=t2&&v(j)>vr(i)
      y(i,j)=t2*vr(i)+t1*t2*a1-0.5*a1*t1^2;
    else if v2(i,j)/a2<t2&&v(j)>vr(i)
        y(i,j)=t2*v(j)-0.5*(t2-t1)^2-v2(i,j)^2/(2*a2);
      else
          y(i,j)=0;
      end
    end
  end
end
surf(v,vr,y); box on
shading interp;
title('Honda Algorithm')
xlabel('v/(m·s-1)')
ylabel('vr/(m·s-1)')
zlabel('db/m')
```

various tests which comply with Chinese regulations and complex road conditions, and allow the performance of AEB systems to be compared in typical rear-end collisions scenarios.

(1) Test Scenarios

In order to compare the various functions of the AEB system, test scenarios need to be defined based on major sources of information describing real-world rear-end crashes on the road.

For the collision accidents in a city, a common type of accident that causes driver injury is a low-speed rear-end collision. An accident occurs when a driver in the following vehicle is distracted and does not notice that the vehicle in front has stopped. Such an accident could easily have caused neck injuries to the passengers in both vehicles. In this case, the test vehicle is

Table 6.3: AEB test scenarios

C-NCAP		Illustration	Test description
CCRs	VT is stationary.	20 km/h, 30 km/h, 40 km/h 0 km/h	Approaching a stopped vehicle at test speeds from 20–30 km/h in 5 km/h increments.
CCRm	VT is moving slowly at constant speed.	30 km/h, 45 km/h, 65 km/h 20 km/h	Approaching a moving target at 20 km/h. The speed of the test vehicle is 30 km/h, 45 km/h , and 65 km/h.
CCRb	VT and VUT are traveling at same speed and maintain constant speed, then lead vehicle applies brake.	12 m or 40 m 50 km/h 50 km/h 4 m/s²	Approaching a decelerating target, both vehicles initially moving at 50 km/h. Target car has 2 headway conditions (short 12 m and long 40 m) and needs to decelerate to 4 m/s² in 1 s

asked to drive towards the back of a "dummy" target vehicle at speeds between 10 and 50 km/h in 2 scenarios: with the centerlines of the vehicle and target aligned; and with the centerline of the test vehicle offset to the left or to the right of the target. Therefore, AEB City systems are tested across a wide range of speeds and vehicle overlaps.

AEB Interurban systems help the driver avoid a rear-end crash by warning and supporting adequate braking or ultimately stopping the vehicle by themselves. Car-to-car rear impacts are among the most frequent accidents on Chinese roads.

For AEB systems, C-NCAP evaluates the automatic brake function and the forward collision warning function in three different driving scenarios: car-to-car rear stationary (CCRs), car-to-car rear moving (CCRm), and car-to-car rear braking (CCRb), as shown in Table 6.3.

The CCRs was given an additional speed range for approaches at 20–40 km/h, and it increases by 10 km/h. The CCRm test scenario for a moving target was defined as a target moving

Table 6.4: AEB test scenarios comparison between Euro NCAP and C-NCAP

Test Items	Test Scenarios					
	CCRs		CCRm		CCRb	
	Euro NCAP	C-NCAP	Euro NCAP	C-NCAP	Euro NCAP	C-NCAP
Vehicle Speed	10–50 km/h with increments of 5 km/h	20 km/h	30–80 km/h with increments of 5 km/h	30 km/h	50 km/h (12 m, -2 and -6 m/s^2)	50 km/h (12 m, -4 m/s^2)
		30 km/h		45 km/h		
		40 km/h		65 km/h	50 km/h (40 m, -2 and -6 m/s^2)	50 km/h (40 m, -4 m/s^2)

at 20 km/h, with approach speeds 30, 45, and 65 km/h, and these speeds were similarly drawn from the accidents study. The CCRb test represents a braking (decelerating) target car. Both test and target vehicles are moving at 50 km/h. The tests are devised to represent a two headway conditions: a long headway of 40 m, and a short headway of 12 m typical of the following distance in busy traffic; and the deceleration of VT should reach -4 m/s^2 within 1 s and error should not exceed ±0.25 m/s^2 until the end of test. The difference in test scenarios between Euro NCAP (3.0.2 version) and C-NCAP (2018 edition) are detailed in Table 6.4.

(2) Test Procedure

After defining the scenarios that reflect real-world accident situations and identifying realistic and practical test objectives, the next step is to define the detail of the test procedure. The aim of the procedure is to provide accurate and repeatable results while minimizing the test burden. Therefore, the procedure starts with the lowest test speed specified for the particular scenario. Test speed was then increased in 10 km/h increments until a test speed is reached where the AEB system no longer avoids the collision and a collision occurs between the test vehicle and target vehicle. At this stage, the test is repeated at a speed 5 km/h lower than that in which the collision occurs. The performance of AEB system is measured in all test scenarios. For the test scenarios CCRs, CCRm, and CCRb, an additional assessment of the vehicle FCW system (if present) was also undertaken. The process for determining the tests to be undertaken is shown in Fig. 6.9.

(3) AEB Tests in CarSim Software

The C-NCAP performance tests aim to evaluate the effectiveness of the AEB system under different driving conditions. According to C-NCAP Management Regulation (2018 Edition), three different driving scenarios (CCRs, CCRm, and CCRb) are selected and tested in CarSim software, respectively. CarSim is one of the commonly used vehicle dynamics simulation soft-

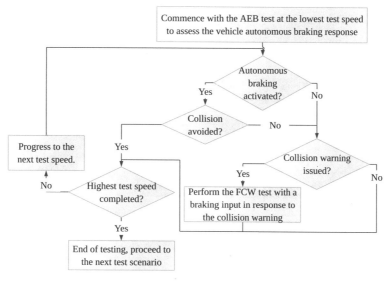

Figure 6.9: Flow diagram of AEB testing.

ware of real-world validation by automotive engineers. The simulation speed is 3–6 times faster than the actual operation, and it can provide an accurate vehicle model to simulate the actual driving scene of the vehicle.

The test results are shown in Fig. 6.10. For the CCRs test scenario, the test vehicle approaches a stopped vehicle at test speeds from 20–40 km/h. VUT speed starts at 20 km/h and increases in 10 km/h increments up to 40 km/h. The initial distance between the vehicles is 100 m, as shown in Fig. 6.10a. For the CCRm test scenario, the test vehicle approaches a moving target at a constant speed of 20 km/h, and the VUT speeds are set to 30, 45, and 65 km/h, respectively. Likewise, the initial distance between the vehicles is 100 m, as shown in Fig. 6.10b. For the CCRb test, it can be divided into two scenarios based on the different headway conditions. VUT and VT are at the same speed of 50 km/h along the planned route, headway should be, respectively 12 m and 40 m, as shown in Fig. 6.10c, d. The deceleration of VT should reach 4 m/s² within 1 s.

6.3 MULTIPLE-VEHICLE COLLISION AVOIDANCE USING V2X COMMUNICATION

In an emergency braking situation, a driver usually determines his or her braking action by observing the brake lights of the vehicle immediately ahead. However, in a heavy traffic situation, if several vehicles in front of the driver make emergency braking one after another, even if the driver in each vehicle immediately takes braking action, the vehicle behind may still have no time to avoid a collision. In addition, the driver's reaction time over all the vehicles ahead will

(a) CCRs

(b) CCRm

(c) CCRb (12 m)

(d) CCRb (40 m)

Figure 6.10: **AEB** tests in CarSim software.

accumulate with the increase of the number of vehicles, and eventually further exacerbate the collision situation. The driver's reaction time usually refers to the time for a driver from seeing the brake light of the vehicle ahead to actually stepping on the brake pedal it is generally between 0.75 s and 1.5 s. As a result, a single emergency event can often lead to multi-vehicle chain accidents, as shown in Fig. 6.11.

Such chain collisions can be avoided or their severity can be lessened by reducing the delay between the time of an emergency event and the time at which the vehicles behind are informed about it. One way to accomplish this task is to use the V2X Communication (i.e., V2V: Vehicle-to-Vehicle or V2I: Vehicle-to-Infrastructure) based wireless networking technology for propagating such road emergency information.

The objective of this section is to use the V2V communication for developing Cooperative Collision Avoidance System (CCAS) which can alleviate the chain-collision accidents as described above. The basic idea of CCAS is as follows: when a vehicle meets an emergency event,

Figure 6.11: Multi-vehicle collision scenarios.

it generates a wireless collision warning message which is forwarded to the vehicles behind in a multi-hop fashion. Since the delay time of wireless message transmission is far less than the cumulative drivers' reaction time, CCAS can deliver warning messages to vehicles behind way before they actually know about it through the brake lights of the vehicles ahead. This way, a driver can be given significantly more time to react to emergency events that are out of their visual range. As a result, the number of vehicles that get into chain collisions can be reduced. In order to design an effective and reliable CCAS based on V2X communication, the communication structures, spacing policy, and string stability that affect the performance of multi-vehicle collision avoidance are described, respectively.

6.3.1   COMMUNICATION STRUCTURES

There are many possible communication graphs, which specify the flow of information between vehicles (some of them are shown in Fig. 6.12). The communication structure is an important factor, which, in particular, influences string stability and the active safety of vehicular platoon.

The practical implementation of the platooning algorithms requires them to be decentralized, that is each vehicle in a platoon gathers the information from neighboring or leading vehicles and decides on its own what should be the vehicle's response to the current road situations.

It is important to mention that not every method of information flow topology is suitable to use in a real traffic scenario. For instance, each vehicle receives data from every vehicle in the platoon may be difficult to implement into the CCAS with many vehicles, since such systems require too many communication channels and are too complex. In this chapter, the control algorithm is based on a unidirectional information flow topology (Fig. 6.12c), therefore the vehicle behind receives the data from the vehicle ahead and the leading vehicle. Such a solution is considered to be simple and effective, since the vehicles may be always in the range of the

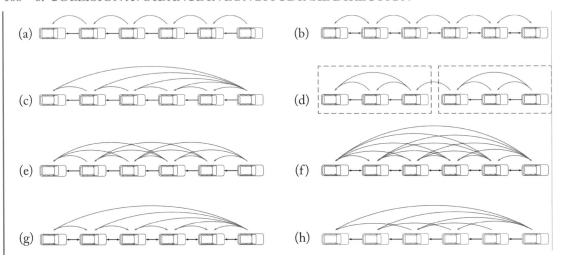

Figure 6.12: Information flow topologies.

wireless network. Another important feature of this kind of unidirectional information flow topology is the fact that it can improve string stability with small inter-vehicle distances.

In this chapter, the control algorithm is based on a unidirectional information flow topology (Fig. 6.12c). Blue arrows indicate wireless communication, while radars or other ranging sensor are denoted with black arrows. Therefore, the vehicle behind receives the data from the vehicle ahead and the leading vehicle. Such a solution is considered to be simple and effective, since the vehicles may be always in the range of the wireless network. Another important feature of this kind of unidirectional information flow topology is the fact that it can improve string stability with small inter-vehicle distances.

6.3.2   SPACING POLICY

Another important factor, which has a great influence on traffic flow and string stability in vehicle platoons is a spacing policy, which defines the distance between consecutive vehicles, and is the basis for the vehicle platoon control algorithm implementation (providing a reference distance input value for it). The ideal situation for the spacing policy is that the distance between all adjacent vehicles traveling in the same direction of the road is small and stable, but combined with the actual operation of the vehicle, a too small distance between vehicles is likely to cause traffic accidents. However an excessively large inter-vehicle distance not only reduces the traffic capacity, but also may increase cut-in vehicles, thereby affecting the safety of vehicle platoon. Therefore, the spacing policy will directly determine the safety and efficiency of the vehicle platoons.

The spacing policy is mainly divided into two types: constant spacing and variable spacing. The constant spacing policy leads to a higher throughput, because the distance between vehicles is independent of the vehicles' velocity. The spacing error with a constant spacing policy is expressed in the following way:

$$D_i(t) = X_{i-1}(t) - X_i(t) - X_{r,i} - L_i, \tag{6.8}$$

where X_{i-1} and X_i indicate the absolute position of the two consecutive vehicles, $X_{r,i}$ is a desired constant spacing distance, and L_i is the length of the ith vehicle.

However, the constant spacing strategy poses a great challenge to the selection of the spacing value. At the same time, the adoption of this strategy requires a more complicated communication topology, and in the case of external interference or large communication delay, this strategy will affect the stability and safety of the platoon [90].

The variable spacing policy is believed to improve string stability and safety of the traffic flow. Among the variable spacing policies, the representatives are mainly the safety spacing policies based on the time headway, which are divided into the constant time headway (CTH) policy and the variable time headway (VTH) policy. The time headway refers to the time interval between the front bumpers of two adjacent vehicles in the platoon passing a certain section. For the variable spacing policy, it is expected that the inter-vehicle distance can be adjusted according to the driving environment. Compared with the constant spacing policy, the variable spacing policy can adapt to a more complex driving environment.

As shown in Fig. 6.13, the desired relative position with CTH policy is given by

$$x_{r,i}(t) = R_i + H_i V_i(t), \tag{6.9}$$

where, R_i stands for desired distance at standstill, H_i is a desired headway time, and $V_i(t)$ is a velocity of the ith vehicle.

It can be seen from Equation (6.9) that the desired distance is proportional to the speed of the host vehicle. When the vehicle speed increases, the corresponding desired spacing also increases. This is because once the emergency braking occurs in the preceding vehicle, the higher the speed of the following vehicle, the more braking distance is needed to avoid collision.

The constant headway policy with the headway time $H_i = 0$ becomes a constant spacing policy.

Unlike the CTH policy, the VTH policy considers that the headway time H_i in Equation (6.9) no longer remains constant, but varies with the surrounding driving environment. Yanakiev and Kanellakopoulos [91] believe that t the headway time H_i is not only related to the speed of the host vehicle, but also to the speed of the preceding vehicle. When the speed of the preceding vehicle is greater than the speed of the following vehicle, the headway time H_i can be appropriately reduced to improve traffic capacity; when the speed of the preceding vehicle is less than the speed of the following vehicle, the headway time H_i must be increased to ensure driving safety. The headway time H_i can be expressed as:

$$H_i = H_0 - C_v V_{rel}, \tag{6.10}$$

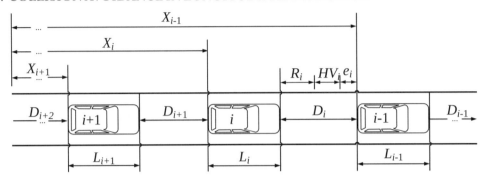

Figure 6.13: Schematic overview of a vehicular platoon.

where H_0 and C_v are constants greater than 0 and V_{rel} is the relative speed of two adjacent vehicles.

It can be seen from Equation (6.10) that the headway time is related to the relative speed of the two consecutive vehicles, and decreases as the relative speed increases.

6.3.3   STRING STABILITY

The notion of string stability is concerned with the propagation of the disturbances in up-stream direction of the platoon. In the case of the string unstable platoon, the oscillatory behavior of the leading vehicle may result in so-called "harmonica effect," which may lead to traffic jams or collisions.

A frequently adopted description of string stability in literature is a frequency domain approach, called the performance-oriented approach. In this method the amplification of either distance, velocity, or acceleration error is measured with the transfer function, for example, from Laplace transform of "input" position error $E_{i-1}(s)$ to the Laplace transform of "output" position error $E_i(s)$, where s is a complex variable in frequency domain:

$$\Gamma_i(s) = \frac{E_i(s)}{E_{i-1}(s)}. \qquad (6.11)$$

In case of homogeneous platoon, a transfer function (6.11) is identical for each vehicle in a string $\Gamma_i(s) = \Gamma(s)$. In order to assure that the distance error does not amplify among the vehicles, the $H\infty$-norm of the transfer function (6.11) has to satisfy the following condition:

$$\|\Gamma_i(j\omega)\| H_\infty \leq 1, \quad 1 \leq i \leq m. \qquad (6.12)$$

A frequency domain is not always suitable, for instance when the platoon is described by a nonlinear system. Then a time-domain approach can be used. It is required that the peak in distance error of each consecutive vehicle is smaller or equal to the error of the predecessor in order to maintain string stability for a platoon of N vehicles (6.13)

$$\|e_1\|_\infty \le \|e_2\|_\infty \cdots \le \|e_n\|_\infty .$$ (6.13)

In this work, we seek for string stability in the following stronger sense:

$$|e_{i+1}(t)| \le |e_i(t)| .$$ (6.14)

Such a definition of string stability requires that the distance error does not increase as it propagates through the platoon.

6.3.4   MATHEMATICAL MODEL FOR COOPERATIVE COLLISION AVOIDANCE

Consider a string of vehicles driving on a single-lane highway. We assume that at $t = 0$, they are all driving with equal speeds and equal inter-vehicle distance. The string of vehicles is modeled as an inter-connected system, with each vehicle driver system forming one element of the inter-connected system. The acceleration response of the driver of the ith vehicle is modeled by the following equation:

$$\frac{dv_i(t)}{dt} = K_1 [S_{i-1,i}(t-\tau) - Tv_i(t-\tau)] + K_2 [v_{i-1}(t-\tau) - v_i(t-\tau)]$$ (6.15)

$$\frac{dS_{i-1,i}(t)}{dt} = v_{i-1}(t) - v_i(t),$$ (6.16)

where $v_i(t)$ indicates the velocity of the ith vehicle, and $S_{i-1,i}(t) = X_{i-1} - X_i$ represents the inter-vehicle distance between the ith and $(i-1)$th vehicles, with vehicle 1 being the leading vehicle. τ indicates the response delay of each vehicle-driver system or transmission delay of vehicle. K_1 is the sensitivity to the difference between the desired inter-vehicle distance and the real inter-vehicle distance, whereas K_2 represents the sensitivity of each vehicle to the velocity difference between his/her car and the car immediately ahead. The desired inter-vehicle distance of each vehicle (to the vehicle ahead) is proportional to vehicles' velocity, with the proportionality constant being T (the time headway). We work with a simplified model, in which we assume all the drivers to possess identical dynamics.

In order to predict the future behavior of the vehicles in mixed traffic flow at each time step, a discrete model to predict the dynamic process of collision is derived from Equations (6.15) and (6.16).

Here, we assume that the current time is kT, and take t as the sampling interval of the discrete model, the velocity of following vehicles in traffic flow at $(k+1)T$ are calculated sequentially by continuing iteration of Equation (6.15):

$$v_1[(k+1)T] = v_1(kT) + \tau * a(kT)$$

$$\vdots$$

$$v_3\left[(k+1)\,T\right] = v_3\left(kT\right) + \tau * \left[K_1\left(S_2\left(kT\right) - T v_3\left(kT\right)\right) + K_2\left(v_2\left(kT\right) - v_3\left(kT\right)\right)\right]$$
$$= \tau K_2 v_2\left(kT\right) + \left(1 - \tau K_1 T - \tau K_2\right) v_3\left(kT\right) + \tau K_1 S_2\left(kT\right)$$

$$\vdots$$

$$v_n\left[(k+1)\,T\right] = v_n\left(kT\right) + \tau * \left[K_1\left(S_{n-1}\left(kT\right) - T v_n\left(kT\right)\right) + K_2\left(v_{n-1}\left(kT\right) - v_n\left(kT\right)\right)\right]$$
$$= \tau K_2 v_{n-1}\left(kT\right) + \left(1 - \tau K_1 T - \tau K_2\right) v_n\left(kT\right) + \tau K_1 S_{n-1}\left(kT\right).$$

In the analysis of the dynamic performance of mixed traffic flow, the inter-vehicle spacing at each state, is another important characteristic parameter. Transfer functions describing the propagation of spacing errors are generated by using the position of each vehicle as a state. Therefore, we will derive all the inter-vehicle spacing of traffic flow at $(k+1)\,T$ using the following transformation:

$$S_1\left[(k+1)\,T\right] = S_1\left(kT\right) + \tau * \left[v_1\left(kT\right) - v_2\left(kT\right)\right]$$

$$\vdots$$

$$S_2\left[(k+1)\,T\right] = S_2\left(kT\right) + \tau * \left[v_2\left(kT\right) - v_3\left(kT\right)\right] \tag{6.17}$$

$$\vdots$$

$$S_{n-1}\left[(k+1)\,T\right] = S_{n-1}\left(kT\right) + \tau * \left[v_{n-1}\left(kT\right) - v_n\left(kT\right)\right].$$

We assume $a_1 = \tau K_2, a_2 = 1 - \tau K_1 T - \tau K_2, a_3 = \tau K_1$, then define the state vector and predicted outputs for the predictive state-space model as:

$$X\left(kT\right) = \begin{bmatrix} v_1\left(kT\right) \\ v_2\left(kT\right) \\ \vdots \\ v_n\left(kT\right) \\ S_1\left(kT\right) \\ S_2\left(kT\right) \\ \vdots \\ S_{n-1}\left(kT\right) \end{bmatrix}. \tag{6.18}$$

Rearranging the definitions of v and S given in Equations (6.17) and (6.18), the discrete state space model is given by:

$$X\left[(k+1)\,T\right] = AX\left(kT\right) + Ba\left(kT\right), \tag{6.19}$$

where

$$A_{(2n-1)*(2n-1)} = \begin{bmatrix} A_{11} & A_{12} \\ A_{21} & A_{22} \end{bmatrix}, \quad B = \begin{bmatrix} 1 \\ 0 \\ \vdots \\ 0 \end{bmatrix}_{n*1}$$

$$A_{11} = \begin{bmatrix} 1 & 0 & 0 & \cdots & 0 \\ a_1^1 & a_2^2 & 0 & \cdots & 0 \\ 0 & a_1^2 & a_2^3 & \cdots & 0 \\ \vdots & \vdots & \vdots & \ddots & 0 \\ 0 & 0 & 0 & a_1^{n-1} & a_2^n \end{bmatrix}_{n*n}, A_{12} = \begin{bmatrix} 0 & 0 & 0 & \cdots & 0 \\ \tau K_1 & 0 & 0 & \cdots & 0 \\ 0 & \tau K_1 & 0 & \cdots & 0 \\ \vdots & \vdots & \vdots & \ddots & 0 \\ 0 & 0 & 0 & 0 & \tau K_1 \end{bmatrix}_{n*n-1}$$

$$A_{21} = \begin{bmatrix} \tau & -\tau & 0 & \cdots & 0 \\ 0 & \tau & -\tau & \cdots & 0 \\ 0 & 0 & \tau & \cdots & 0 \\ \vdots & \vdots & \vdots & \ddots & 0 \\ 0 & 0 & 0 & \tau & -\tau \end{bmatrix}_{n-1*n}, A_{22} = \begin{bmatrix} 1 & 0 & 0 & \cdots & 0 \\ 0 & 1 & 0 & \cdots & 0 \\ 0 & 0 & 1 & \cdots & 0 \\ \vdots & \vdots & \vdots & \ddots & 0 \\ 0 & 0 & 0 & 0 & 1 \end{bmatrix}_{n-1*n-1}.$$

6.3.5   NUMERICAL SIMULATIONS

Consider the following scenario: a leading vehicle in a platoon is initially travelling at a typical speed of about 30 m/s, (i.e., 108 km/h), with an inter-vehicle spacing of 36 m, (i.e., $T = 1.2$ s). At $t = 5$ s, the leading vehicle begins to execute an abrupt deceleration of -5 m/s^2, and decelerates continuously for 5 s. According to the statistical results of driving behavior, this value of τ is selected as 0.6 s, however, the values of K_1 and K_2 need to be determined to ensure stable, non-oscillatory behavior for vehicular platoon. We now present simulations showing the effect of the leading vehicle's deceleration on the vehicles behind, when information of this deceleration is transmitted in each of the modes demonstrated in Fig. 6.12 of Section 6.3.1.

As discussed above, for a vehicle platoon consisting entirely of human-driven vehicles, the cumulative driver reaction time will result in a pile-up accident, as shown in Fig. 6.14. Refer to Fig. 6.14a, which shows the velocity profiles of 12 vehicles, it can be seen that the deceleration of the leading vehicle is passed from the front of the platoon to the back, as in Fig. 6.12a. Figure 6.14b shows that the values of the minimum inter vehicle distance keep decreasing with increasing vehicle index, until vehicle 3 is rear-ended by vehicle 4, and crashes occur for all the vehicles behind. As you can see from the figures, if there are more vehicles behind vehicle 12, they will also collide and cause a pile-up. This is the phenomenon of string instability.

The main reason lead to pile-up accidents is that the time headway between all the vehicles in the platoon is too short, the information of deceleration of the leading vehicle is transmitted to the following vehicles through the brake light, which is too slow to give the driver enough time to react to avoid the impending collision. It can also be inferred that the shorter the time headway maintained by a vehicle (to the one ahead of it), the more likely the onset of string instability; at higher time headways, string instability does not manifest.

Now, consider a scenario when all 12 vehicles are equipped with the CCAS, and the leading vehicle performs the same deceleration profile as in the previous example. In this case, at the moment when the leading vehicle begins to slow down, all the vehicles behind get the deceleration information of the leading vehicle almost at the same time, and these vehicles take action

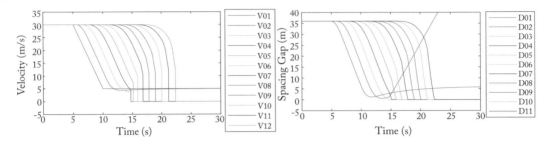

Figure 6.14: **All vehicles unequipped.**

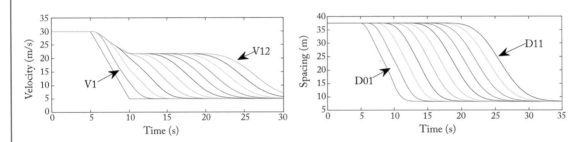

Figure 6.15: **All vehicles equipped.**

about τ seconds after the leading vehicle begins to decelerate, as opposed to 11τ seconds that it would otherwise have taken, if all vehicles were unequipped. The information transmission mode of this type of vehicle platoon is shown in Fig. 6.12c of Section 6.3.1. In addition, once the ICVs get the deceleration information from the leading vehicle, the control system of ICV can immediately operate the braking system until the vehicle stops, and the execution lag time is obviously less than the driver's operation time. Furthermore, the ICVs attempt to increase their time headway to the vehicle immediately ahead (from the original $T = 1.2$ s to $T = 1.65$ s), in anticipation of the imminent slowdown, the results are shown in Fig. 6.15. The trend of decreasing values of the minimum vehicle velocity and minimum inter-vehicle distance (with increasing vehicle index) is no longer seen. This would be true even for any arbitrary number of vehicles behind vehicle 12, if they too were equipped.

6.4   SUMMARY

This chapter discusses longitudinal collision avoidance technologies address front-to-rear accidents, describing the characteristics and the functions of the FCW and AEB systems currently on the market. In addition, the cooperative collision avoidance system is introduced, and its application in collision avoidance of pile-up accidents is simulated and analyzed.

Many previous studies have provided the effectiveness of FCW and AEB technologies or integrated systems based on field tests or simulations. In this chapter, two main types of algorithms, kinematic and perceptual approaches, are introduced and compared by theoretical analysis and numerical simulations. Next, we introduce the three different scenarios and procedures in the AEB car-to-car test, which are part of the European New Car Assessment Program (Euro NCAP) and China New Car Assessment Program (C-NCAP). For the problem of safety of multi-vehicle system, we analyzed the causes of pile-up accident of all human driven vehicles, and concerned the alleviation of pile-up crashes on the highway—with the equipped vehicles being alerted to the possibility of the occurrence of a pile-up crash, by means of a cooperative collision avoidance system. Finally, the simulation results under the two scenarios are analyzed and compared.

CHAPTER 7

Conclusions and Future Works

7.1 CONCLUSIONS

Different path-planning and tracking algorithms for vehicle collision avoidance in lateral and longitudinal motion directions are discussed in this book. Based on the reviews in the previous sections, it can be deduced that a successful collision avoidance demands a complex system which considers various factors such as calculation costs, constraints (vehicle dynamics, obstacles, and environment), as well as the incorporation of latest computational and sensing devices.

For the path planning strategy, the selected approach should be able to provide flexible global routes to the destination for the host vehicle during navigation. In addition, it should provide an emergency trajectory which does not collide with the obstacles, possess small usage of memory space and space range during the planning action. It also should include the consideration of moving obstacles, consider the risk of lane departure after collision avoidance, and be able to operate in complex high nonlinearity scenarios.

Path-tracking strategies will then act as a path following controller to enable the vehicle to navigate through the feasible planned trajectories. To act as reliable automated motion guidance, the path tracking should consist of a controller which considers the nonlinearity of the vehicle model. Furthermore, it should have fast computational time to provide multi-actuator intervention during collision avoidance navigation. Overall, the combination of good path-planning and tracking strategies will enable a robust collision avoidance system in various traffic scenarios.

Many recent studies have focused on FCW and AEB to improve the traffic safety. However, most previous research on collision avoidance and road safety simulation considered only the nearest two vehicles. With increasing computing and wireless communication abilities, the CCAS have the potential to further mitigate the crash accidents in longitudinal direction. The CCAS has a great potential for avoiding multi-vehicle accidents or, at least, for minimizing the impact if the collision is unavoidable by simultaneously controlling the braking of multiple vehicles.

Collision avoidance technologies in lateral and longitudinal motion directions have made great progress both academically and commercially in the past ten years. However, the vehicle is an object that needs to be studied as a whole. Therefore, the future collision avoidance technology should be able to control both lateral and longitudinal movement at the same time, so as to further improve the traffic safety and ride comfort of vehicle.

7.2   FUTURE WORKS

Alhough the industry is showing much progress in the area of collision avoidance for intelligent connected vehicles, but the complexity of road traffic in the real world is rapidly increasing day by day. This prediction gives the picture that collision avoidance, as a part of ADAS, still need to be enhanced with the help of today's autonomous driving and V2X communications technologies.

Most major carmakers are actively promoting their progress on developing latest collision avoidance system in recente years. It is important to be noted that an advanced collision avoidance research will eventually contribute toward the implementation of Cooperative Vehicle Infrastructure System (CVIS) a concept which incorporates V2X communication into the transportation and vehicles. Thus, the collaboration between industrial, academia, and governmental sectors is a necessity for the real-time collision avoidance system to be enacted globally. This in return will promise a reduced number of road casualties, and subsequently fewer road fatalities, it may be one of the potential ways to achieve the ultimate goal of zero deaths on road.

APPENDIX A

MATLAB Programs

The MATLAB program in Appendix_1 is used to solve the path-planning problem as shown in Fig. 2.3 in Section 2.1.

Listing A.1: Path-planning problem (*Continues*)

```
%Define the 2-D map array for A* Algorithm
X_max=7;
Y_max=7;
MAP=2*(ones(X_max,Y_max));
axis([1 X_max+1 1 Y_max+1])
grid on
hold on
% Initialize MAP with location of the start, obstacles, and target
T_x=6;
T_y=2;
MAP(T_x,T_y)=0;
plot(T_x+.5,T_y+.5,'gp','MarkerSize',15);
text(T_x-0.2,T_y+1.2,'Target Point','FontSize',14)
hold on
O_x=[2 4 4 4 4];
O_y=[2 2 3 4 5];
for i = 1:length(O_x)
    MAP(O_x(1,i),O_y(1,i))=-1;
end
plot(O_x+.5,O_y+.5,'rs','MarkerSize',15); % Plot the obstacles
text(max(O_x),max(O_y)+1,'Obstacles','FontSize',14)
hold on
S_x=2;
S_y=3;
MAP(S_x,S_y)=1;
plot(S_x+.5,S_y+.5,'bo','MarkerSize',15);
text(S_x-0.2,S_y+1.2,'Initial Point','FontSize',14);
```

Listing A.1: (*Continued*). Path-planning problem (*Continues*)

```
OPEN=[];
CLOSED=[O_x',O_y'];
CLOSED_COUNT=size(CLOSED,1);

xNode=S_x;
yNode=S_y;
OPEN_COUNT=1;
path_cost=0;
goal_distance=sqrt((S_x-T_x)^2 + (S_y-T_y)^2);
OPEN(OPEN_COUNT,:)=insert_open(xNode,yNode,xNode,yNode,path_cost,goal_distance,-
goal_distance);
OPEN(OPEN_COUNT,1)=0;
CLOSED_COUNT=CLOSED_COUNT+1;
CLOSED(CLOSED_COUNT,1)=xNode;
CLOSED(CLOSED_COUNT,2)=yNode;
NoPath=1;

xTarget = T_x;
yTarget = T_y;

while((xNode ~= xTarget || yNode ~= yTarget) && NoPath == 1)
exp_array=expand_array(xNode,yNode,path_cost,xTarget,yTarget,CLOSED,X_max,Y_max);
 exp_count=size(exp_array,1);
for i=1:exp_count
   flag=0;
   for j=1:OPEN_COUNT
     if(exp_array(i,1) == OPEN(j,2) && exp_array(i,2) == OPEN(j,3) )
        OPEN(j,8)=min(OPEN(j,8),exp_array(i,5));
        if OPEN(j,8)== exp_array(i,5)
           OPEN(j,4)=xNode;
           OPEN(j,5)=yNode;
           OPEN(j,6)=exp_array(i,3);
           OPEN(j,7)=exp_array(i,4);
```

Listing A.1: (*Continued*). Path-planning problem (*Continues*)

```
   end
        flag=1;
      end
   end
if flag == 0
      OPEN_COUNT = OPEN_COUNT+1;
OPEN(OPEN_COUNT,:)=insert_open(exp_array(i,1),exp_array(i,2),xNode,yNode,exp_ar-
ray(i,3),exp_array(i,4),exp_array(i,5));
   end
 end
 index_min_node = min_fn(OPEN,OPEN_COUNT,xTarget,yTarget);
 if (index_min_node ~= -1)
  xNode=OPEN(index_min_node,2);
  yNode=OPEN(index_min_node,3);
  path_cost=OPEN(index_min_node,6);
 CLOSED_COUNT=CLOSED_COUNT+1;
 CLOSED(CLOSED_COUNT,1)=xNode;
 CLOSED(CLOSED_COUNT,2)=yNode;
 OPEN(index_min_node,1)=0;
 else
    NoPath=0;
 end;
end;
i=size(CLOSED,1);
Optimal_path=[];
xval=CLOSED(i,1);
yval=CLOSED(i,2);
i=1;
Optimal_path(i,1)=xval;
Optimal_path(i,2)=yval;
i=i+1;
inode=0;
  %Traverse OPEN and determine the parent nodes
  parent_x=OPEN(node_index(OPEN,xval,yval),4);
  parent_y=OPEN(node_index(OPEN,xval,yval),5);
```

Listing A.1: (*Continued*). Path-planning problem

```
xStart=S_x;
yStart=S_y;
  while( parent_x ~= xStart || parent_y ~= yStart)
      Optimal_path(i,1) = parent_x;
Optimal_path(i,2) = parent_y;
      inode=node_index(OPEN,parent_x,parent_y);
      parent_x=OPEN(inode,4);
      parent_y=OPEN(inode,5);
      i=i+1;
  end
 plot(Optimal_path(:,1)+.5,Optimal_path(:,2)+.5);
```

This MATLAB program is a model predictive control S function used to solve the path tracking problem, which can be called by the co-simulation shown in Figure 3.11 in Section 3.3.

Listing A.2: Model predictive control function (*Continues*)

```
function [sys,x0,str,ts]=MPCController1(t,x,u,flag)
switch flag
  case 0
    [sys,x0,str,ts]=mdlInitializeSizes;
  case 2
    sys=mdlUpdates(t,x,u);
  case 3
    sys=mdlOutputs(t,x,u);
  case {1,4,9}
    sys=[];
  otherwise
    error(['unhandled flag=',num2str(flag)]);
end
function[sys,x0,str,ts]=mdlInitializeSizes
sizes=simsizes;
sizes.NumContStates =0;
sizes.NumDiscStates =5;
sizes.NumOutputs =1;
sizes.NumInputs =7;
sizes.DirFeedthrough=1;
sizes.NumSampleTimes=1;
sys=simsizes(sizes);
x0=[0.0001;0.0001;0.00001;0.00001;0.00001];
global U;
U=[0];
str=[]; %set str to an empty matrix
ts=[0.01 0];%sample time :[period ,offset]
function sys=mdlUpdates(t,x,u)
```

Listing A.2: (*Continued*). Model predictive control function (*Continues*)

```
sys=x;
function sys=mdlOutputs(t,x,u)
global a b;
global U;
tic
Nx=5;
Nu=1;
Ny=2;
Np=20;
Nc=10;
T=0.01;
y_dot=u(1)/3.6;
phi=u(2)*3.141592654/180;
phi_dot=u(3)*3.141592654/180;
Y=u(4);
X=u(5);
global x_dot
x_dot=20;
lf=1.015;lr=1.895;
Ccf=66900;Ccr=62700;
m=1270;g=9.8;I=1537.6;
X_predict=zeros(Np,1);
phi_ref=zeros(Np,1);
Y_ref=zeros(Np,1);
kesi=zeros(Nx+Nu,1);
kesi(1)=y_dot;%
kesi(2)=phi;
kesi(3)=phi_dot;
kesi(4)=Y;
kesi(5)=X;
kesi(6)=U(1);
delta_f=U(1);
u_piao=zeros(Nx,Nu);
Q_cell=cell(Np,Np);
for i=1:1:Np
```

Listing A.2: (*Continued*). Model predictive control function (*Continues*)

```
for j=1:1:Np
    if i==j
        Q_cell{i,j}=[2000 0;0 10000;];
    else
Q_cell{i,j}=zeros(Ny,Ny);
    end
  end
end
R=5*10^5*eye(Nu*Nc);
Row=1000;
a=[1 - (1296*T)/127, 0,-(203087*T)/12700, 0, 0;
0,1,T,0,0;
(50913*T)/15367,0,1 - (7351983*T)/384175,0,0;
T*cos(phi),T*(20*cos(phi)-y_dot*sin(phi)),0,1,0;
-T*sin(phi),-T*(20*sin(phi)+y_dot*cos(phi)),0,0,1];
b=[(13380*T)/127;
 0;
 (1358070*T)/15367;
 0;
 0];
A_cell=cell(2,2);
B_cell=cell(2,1);
A_cell{1,1}=a;
A_cell{1,2}=b;
A_cell{2,1}=zeros(Nu,Nx);
A_cell{2,2}=eye(Nu);
B_cell{1,1}=b;
B_cell{2,1}=eye(Nu);
A=cell2mat(A_cell);
B=cell2mat(B_cell);
C=[0 1 0 0 0 0 ;0 0 0 1 0 0];
d_k=zeros(Nx,1);
state_k1=zeros(Nx,1);
state_k1(1,1)=y_dot+T*(-x_dot*phi_dot+2*(Ccf*(delta_f-(y_dot+lf*phi_dot)/x_dot)+Ccr*(l-
r*phi_dot-y_dot)/x_dot)/m);
```

Listing A.2: (*Continued*). Model predictive control function (*Continues*)

```
state_k1(2,1)=phi+T*phi_dot;
state_k1(3,1)=phi_dot+T*((2*lf*Ccf*(delta_f-(y_dot+lf*phi_dot)/x_dot)-2*lr*Ccr*(lr*phi_
dot-y_dot)/x_dot)/I);
state_k1(4,1)=Y+T*(x_dot*sin(phi)+y_dot*cos(phi));
state_k1(5,1)=X+T*(x_dot*cos(phi)-y_dot*sin(phi));
d_k=state_k1-a*kesi(1:5,1)-b*kesi(6,1);
d_piao_k=zeros(Nx+Nu,1);
d_piao_k(1:5,1)=d_k;
d_piao_k(6,1)=0;
PSI_cell=cell(Np,1);
THETA_cell=cell(Np,Nc);
GAMMA_cell=cell(Np,Np);
PHI_cell=cell(Np,1);
for p=1:1:Np
  PHI_cell{p,1}=d_piao_k;
  for q=1:1:Np
    if q<=p
      GAMMA_cell{p,q}=C*A^(p-q);
    else
      GAMMA_cell{p,q}=zeros(Ny,Nx+Nu);
    end
  end
end
for j=1:1:Np
  PSI_cell{j,1}=C*A^j;
  for k=1:1:Nc
    if k<=j
      THETA_cell{j,k}=C*A^(j-k)*B;
    else
      THETA_cell{j,k}=zeros(Ny,Nu);
    end
  end
end
PSI=cell2mat(PSI_cell);%size(PSI)=[Ny*Np Nu*Nc]
THETA=cell2mat(THETA_cell);
GAMMA=cell2mat(GAMMA_cell);
```

Listing A.2: (*Continued*). Model predictive control function (*Continues*)

```
PHI=cell2mat(PHI_cell);
Q=cell2mat(Q_cell);
H_cell=cell(2,2);
H_cell{1,1}=2*THETA'*Q*THETA+R;
H_cell{1,2}=zeros(Nu*Nc,1);
H_cell{2,1}=zeros(1,Nu*Nc);
H_cell{2,2}=Row;
H=cell2mat(H_cell);
error_1=zeros(Ny*Np,1);
Yita_ref_cell=cell(Np,1);
T_all=2.5;
global xf;
xf=50;
for p=1:1:Np
  if t+p*T>T_all
     X_predict(Np,1)=X+x_dot*Np*T;
    Y_ref(p,1)=3.75;
    phi_ref(p,1)=0;
     Yita_ref_cell{p,1}=[phi_ref(p,1);Y_ref(p,1)];
  else
    X_predict(p,1)=X+x_dot*p*T;
    Y_ref(p,1)=37.5*(X_predict(p,1)/xf)^3-15*3.75*(X_predict(p,1)/xf)^4+6*3.75*(X_pre-
dict(p,1)/xf)^5;
    phi_ref(p,1)=atan((9*X_predict(Np,1)^4)/25000000 - (9*X_predict(Np,1)^3)/250000 +
(9*X_predict(Np,1)^2)/10000);
     Yita_ref_cell{p,1}=[phi_ref(p,1);Y_ref(p,1)];
  end
end
Yita_ref=cell2mat(Yita_ref_cell);
error_1=Yita_ref-PSI*kesi-GAMMA*PHI;
f_cell=cell(1,2);
f_cell{1,1}=2*error_1'*Q*THETA;
f_cell{1,2}=0;
f=-cell2mat(f_cell);
A_t=zeros(Nc,Nc);
for p=1:1:Nc
```

Listing A.2: (*Continued*). Model predictive control function

```
for q=1:1:Nc
    if q<=p
        A_t(p,q)=1;
else
        A_t(p,q)=0;
    end
    end
end
A_I=kron(A_t,eye(Nu));
Ut=kron(ones(Nc,1),U(1));
umin=-0.2;
umax=0.2;
delta_umin=-0.06;
delta_umax=0.06;
Umin=kron(ones(Nc,1),umin);
Umax=kron(ones(Nc,1),umax);
ycmax=[0.2;4];
ycmin=[-0.2;-4];
Ycmax=kron(ones(Np,1),ycmax);
Ycmin=kron(ones(Np,1),ycmin);
A_cons_cell={A_I zeros(Nu*Nc,1);-A_I zeros(Nu*Nc,1);
THETA zeros(Ny*Np,1);-THETA zeros(Ny*Np,1)};
b_cons_cell={Umax-Ut;-Umin+Ut;
Ycmax-PSI*kesi-GAMMA*PHI;-Ycmin+PSI*kesi+GAMMA*PHI};
A_cons=cell2mat(A_cons_cell);
b_cons=cell2mat(b_cons_cell);
M=10;
delta_Umin=kron(ones(Nc,1),delta_umin);
delta_Umax=kron(ones(Nc,1),delta_umax);
lb=[delta_Umin;0];
ub=[delta_Umax;M];
options=optimset('Algorithm','interior-point-convex');
[X,fval,exitflag]=quadprog(H,f,A_cons,b_cons,[],[],lb,ub,[],options);
u_piao(1)=X(1);
U(1)=kesi(6,1)+u_piao(1);
sys=U;
```

References

[1] World Health Organization, Global status report on road safety 2018. https://www.who.int/news-room/fact-sheets/detail/road-traffic-injuries, 2020. 1

[2] National Bureau of Statistics, Statistics on traffic accidents in China by year (1995–2018), http://data.stats.gov.cn/easyquery.htm?cn=C01&zb=A0S0D02&sj=2019, 2020. 1

[3] Volvo Trucks, Trucks V. European accident research and safety report 2013. https://www.volvogroup.com/content/dam/volvo/volvo-group/markets/global/en-en/about-us/traffic-safety/ART-report-2013.pdf, 2013. 1

[4] Rajaram, V. and Subramanian, S. C. Heavy vehicle collision avoidance control in heterogeneous traffic using varying time headway. *Mechatronics*, (50):328–40, 2018. DOI: 10.1016/j.mechatronics.2017.11.010. 1

[5] Ba, Y., et al. Crash prediction with behavioral and physiological features for advanced vehicle collision avoidance system. *Transportation Research Part C-Emerging Technologies*, 74:22–33, 2017. DOI: 10.1016/j.trc.2016.11.009. 1, 3

[6] Chen, Y., Peng, H., and Grizzle, J. Obstacle avoidance for low-speed autonomous vehicles with barrier function. *IEEE Transactions on Control Systems Technology*, 26(1):194–206, 2017. DOI: 10.1109/tcst.2017.2654063. 2

[7] Alonso-Mora, J., Beardsley, P., and Siegwart, R. Cooperative collision avoidance for nonholonomic robots. *IEEE Transactions on Robotics*, 34(2):404–420, 2018. DOI: 10.1109/tro.2018.2793890. 2

[8] Ji, J., et al. Path planning and tracking for vehicle collision avoidance based on model predictive control with multiconstraints. *IEEE Transactions on Vehicular Technology*, 66(2):952–964, 2017. DOI: 10.1109/tvt.2016.2555853. 3

[9] Shah, J., et al. Autonomous rear-end collision avoidance using an electric power steering system. *Proc. of the Institution of Mechanical Engineering Part D: Journal of Automobile Engineering*, 229(12):1638–1655, 2015. DOI: 10.1177/0954407014567517. 3

[10] National Center for Statistics and Analysis, Analyses of rear-end crashes and near-crashes in the 100-car naturalistic driving study to support rear-signaling countermeasure development (DOT HS 810 846), https://www.nhtsa.gov/sites/nhtsa.dot.gov/files/

analyses20of20rear-end20rashes20and20nearcrashes20dot20hs2081020846.pdf,   2007.
5

[11] Cicchino, J. B. Effectiveness of forward collision warning and autonomous emergency braking systems in reducing front-to-rear crash rates. *Accident Analysis and Prevention*, 99:142–152, 2017. DOI: 10.1016/j.aap.2016.11.009. 6

[12] Wiseman, Y. Efficient embedded computing component for anti-lock braking system. *International Journal of Control and Automation*, 11(12):1–10, 2018. DOI: 10.14257/ijca.2018.11.12.01. 6

[13] National Highway Traffic Safety Administration, Human performance evaluation of light vehicle brake assist systems: Final report (DOT HS 810 846), https://www.nhtsa.gov/sites/Nhtsa.dot.gov/files/811251.pdf, 2010. 6

[14] Cicchino, J. B. Effectiveness of forward collision warning and autonomous emergency braking systems in reducing front-to-rear crash rates. *Accident Analysis and Prevention*, 99:142–152, 2017. DOI: 10.1016/j.aap.2016.11.009. 6

[15] Maurya, S. K. and Choudhary, A. Deep learning based vulnerable road user detection and collision avoidance. *IEEE International Conference on Vehicular Electronics and Safety (ICVES)*, pages 1–6, 2018. DOI: 10.1109/icves.2018.8519504. 7

[16] Edwards, M., Nathanson, A., Carroll, J., et al. Assessment of integrated pedestrian protection systems with autonomous emergency braking (AEB) and passive safety components. *Traffic Injury Prevention*, 16(sup1):S2–S11, 2015. DOI: 10.1080/15389588.2014.1003154. 7

[17] Yu, S. H., Shih, O., Tsai, H. M., et al. Smart automotive lighting for vehicle safety. *IEEE Communications Magazine*, 51(12):50–59, 2013. DOI: 10.1109/mcom.2013.6685757. 7

[18] Cho, H., Kim, G. E., and Kim, B. W. Usability analysis of collision avoidance system in vehicle-to-vehicle communication environment. *Journal of Applied Mathematics*, 2014. DOI: 10.1155/2014/951214. 7

[19] Dey, K. C., Rayamajhi, A., Chowdhury, M., et al. Vehicle-to-vehicle (V2V) and vehicle-to-infrastructure (V2I) communication in a heterogeneous wireless network—performance evaluation. *Transportation Research Part C: Emerging Technologies*, 68:168–184, 2016. DOI: 10.1016/j.trc.2016.03.008. 7

[20] Wang, P. W., Wang, L., Li, Y. H., et al. Improved cooperative collision avoidance (CCA) model considering driver comfort. *International Journal of Automotive Technology*, 16(6):989–996, 2015. DOI: 10.1007/s12239-015-0101-7. 7

[21] Ahn, H. and Del Vecchio, D. Safety verification and control for collision avoidance at road intersections. *IEEE Transactions on Automatic Control*, 63(3):630–642, 2017. DOI: 10.1109/tac.2017.2729661. 7

[22] Chen, S., Hu, J., Shi, Y., et al. Vehicle-to-everything (V2X) services supported by LTE-based systems and 5G. *IEEE Communications Standards Magazine*, 1(2):70–76, 2017. DOI: 10.1109/mcomstd.2017.1700015. 8

[23] Li, Y., Chen, W., Peeta, S., et al. Platoon control of connected multi-vehicle systems under V2X communications: Design and experiments. *IEEE Transactions on Intelligent Transportation Systems*, 2019. DOI: 10.1109/tits.2019.2905039. 8

[24] Hajiloo, R., Abroshan, M., Khajepour, A., et al. Integrated steering and differential braking for emergency collision avoidance in autonomous vehicles. *IEEE Transactions on Intelligent Transportation Systems*, 2020, (inpress). DOI: 10.1109/tits.2020.2984210. 9

[25] Schnelle, S., Wang, J., Jagacinski, R., et al. A feedforward and feedback integrated lateral and longitudinal driver model for personalized advanced driver assistance systems. *Mechatronics*, 50:177–188, 2018. DOI: 10.1016/j.mechatronics.2018.02.007. 9

[26] Cheng, S., Li, L., Guo, H. Q., et al. Longitudinal collision avoidance and lateral stability adaptive control system based on MPC of autonomous vehicles. *IEEE Transactions on Intelligent Transportation Systems*, 2019. DOI: 10.1109/tits.2019.2918176. 9

[27] Brännström, M., Coelingh, E., and Sjöberg, J. Decision-making on when to brake and when to steer to avoid a collision. *International Journal of Vehicle Safety 1*, 7(1):87–106, 2014. DOI: 10.1504/ijvs.2014.058243. 9

[28] Mukhtar, A., Xia, L., and Tang, T. B. Vehicle detection techniques for collision avoidance systems: A review. *IEEE Transactions on Intelligent Transportation Systems*, 16(5):2318–2338, 2015. DOI: 10.1109/tits.2015.2409109. 13

[29] Harper, C. D., Hendrickson, C. T., and Samaras, C. Cost and benefit estimates of partially-automated vehicle collision avoidance technologies. *Accident Analysis and Prevention*, 95:104–115, 2016. DOI: 10.1016/j.aap.2016.06.017. 13

[30] Raja, P. and Pugazhenthi, S. Optimal path planning of mobile robots: A review. *International Journal of Physical Sciences*, 7(9):1314–1320, 2012. DOI: 10.5897/IJPS11.1745. 13

[31] Koubâa, A., Bennaceur, H., Chaari, I., et al. *Robot Path Planning and Cooperation-Foundations, Algorithms and Experimentations*. Springer International Publishing, 2018. 14

[32] Klancar, G., Zdesar, A., Blazic, S., et al. *Wheeled Mobile Robotics: From Fundamentals Towards Autonomous Systems*. Butterworth-Heinemann, 2017. 14

[33] Carbone, G. and Gomez-Bravo, F., (Eds). *Motion and Operation Planning of Robotic Systems: Background and Practical Approaches*. Springer, 2015. DOI: 10.1007/978-3-319-14705-5. 14

[34] Qing, G., Zheng, Z., and Yue, X. Path-planning of automated guided vehicle based on improved Dijkstra algorithm. *29th Chinese Control and Decision Conference (CCDC)*, IEEE, 2017. DOI: 10.1109/ccdc.2017.7978471. 14

[35] Zhang, J. D., Feng, Y. J., Shi, F. F., et al. Vehicle routing in urban areas based on the oil consumption weight-Dijkstra algorithm. *IET Intelligent Transport Systems*, 10(7):495–502, 2016. DOI: 10.1049/iet-its.2015.0168. 14

[36] Qu, Y., Zhang, Y., and Zhang, Y. A global path planning algorithm for fixed-wing UAVs. *Journal of Intelligent and Robotic Systems*, 91(3–4):691–707, 2018. DOI: 10.1007/s10846-017-0729-9. 14

[37] Zhao, Y., Zheng, Z., and Liu, Y. Survey on computational-intelligence-based UAV path planning. *Knowledge-Based Systems*, 158:54–64, 2018. DOI: 10.1016/j.knosys.2018.05.033. 15

[38] Song, R., Liu, Y., and Bucknall, R. Smoothed A* algorithm for practical unmanned surface vehicle path planning. *Applied Ocean Research*, 83:9–20, 2019. DOI: 10.1016/j.apor.2018.12.001. 16

[39] A* search algorithm. https://en.wikipedia.org/wiki/A*_search_algorithm, 2020. 18

[40] Dong, Y., Zhang, Y., and Ai, J. Experimental test of unmanned ground vehicle delivering goods using RRT path planning algorithm. *Unmanned Systems*, 5(01):45–57, 2017. DOI: 10.1142/s2301385017500042. 20

[41] Noreen, I., Khan, A., and Habib, Z. A comparison of RRT, RRT*, and RRT*-smart path planning algorithms. *International Journal of Computer Science and Network Security*, 16(10):20, 2016. 20

[42] Yoon, S., Lee, D., Jung, J., and Shim, D. H. Spline-based RRT* using piecewise continuous collision-checking algorithm for car-like vehicles. *Journal of Intelligent and Robotic Systems*, 90(3–4):537–49, 2018. DOI: 10.1007/s10846-017-0693-4. 20

[43] Rashid, A. T., Ali, A. A., Frasca, M., et al. Path planning with obstacle avoidance based on visibility binary tree algorithm. *Robotics and Autonomous Systems*, 61(12):1440–1449, 2013. DOI: 10.1016/j.robot.2013.07.010. 23

[44] Mohanty, P. K. and Parhi, D. R. Optimal path planning for a mobile robot using cuckoo search algorithm. *Journal of Experimental and Theoretical Artificial Intelligence*, 28(1–2):35–52, 2016. DOI: 10.1080/0952813x.2014.971442. 24

[45] Radmanesh, M., Kumar, M., Guentert, P. H., et al. Overview of path-planning and obstacle avoidance algorithms for UAVs: A comparative study. *Unmanned Systems*, 6(02):95–118, 2018. DOI: 10.1142/s2301385018400022. 24

[46] Choi, Y., Kim, D., Hwang, S., et al. Dual-arm robot motion planning for collision avoidance using B-spline curve. *International Journal of Precision Engineering and Manufacturing*, 18(6):835–843, 2017. DOI: 10.1007/s12541-017-0099-z. 24

[47] Tharwat, A., Elhoseny, M., Hassanien, A. E., et al. Intelligent Bézier curve-based path planning model using Chaotic Particle Swarm Optimization algorithm. *Cluster Computing*, 22(2):4745–4766, 2019. DOI: 10.1007/s10586-018-2360-3. 24

[48] Bakdi, A., Hentout, A., Boutami, H., et al. Optimal path planning and execution for mobile robots using genetic algorithm and adaptive fuzzy-logic control. *Robotics and Autonomous Systems*, 89:95–109, 2017. DOI: 10.1016/j.robot.2016.12.008. 24

[49] Ntousakis, I. A., Nikolos, I. K., and Papageorgiou, M. Optimal vehicle trajectory planning in the context of cooperative merging on highways. *Transportation Research Part C: Emerging Technologies*, 71:464–488, 2016. DOI: 10.1016/j.trc.2016.08.007. 28

[50] Chen, W. C., Chen, C. S., Lee, F. C., et al. High speed blending motion trajectory planning using a predefined absolute accuracy. *The International Journal of Advanced Manufacturing Technology*, pages 1–15, 2019. DOI: 10.1007/s00170-019-03973-y. 30

[51] Ji, J., Khajepour, A., Melek, W. W., et al. Path planning and tracking for vehicle collision avoidance based on model predictive control with multiconstraints. *IEEE Transactions on Vehicular Technology*, 66(2):952–964, 2016. DOI: 10.1109/tvt.2016.2555853. 37

[52] Yu, H., Meier, K., Argyle, M., et al. Cooperative path planning for target tracking in urban environments using unmanned air and ground vehicles. *IEEE/ASME Transactions on Mechatronics*, 20(2):541–552, 2014. DOI: 10.1109/tmech.2014.2301459. 37

[53] Normey-Rico, J. E., Alcalá, I., Gómez-Ortega, J., et al. Mobile robot path tracking using a robust PID controller. *Control Engineering Practice*, 9(11):1209–1214, 2001. DOI: 10.1016/s0967-0661(01)00066-1. 37

[54] Marino, R., Scalzi, S., and Netto, M. Nested PID steering control for lane keeping in autonomous vehicles. *Control Engineering Practice*, 19(12):1459–1467, 2011. DOI: 10.1016/j.conengprac.2011.08.005. 38

[55] Xu, S. and Peng, H. Design, analysis, and experiments of preview path tracking control for autonomous vehicles. *IEEE Transactions on Intelligent Transportation Systems*, 21(1):48–58, 2019. DOI: 10.1109/tits.2019.2892926. 40

[56] Zhang, X. and Zhu, X. Autonomous path tracking control of intelligent electric vehicles based on lane detection and optimal preview method. *Expert Systems with Applications*, 121:38–48, 2019. DOI: 10.1016/j.eswa.2018.12.005. 40

[57] Kondo, M. Directional stability (when steering is added). *Journal of the Society of Automotive Engineers of Japan (JSAE)*, 7(5–6):9, 1953. 41

[58] Bai, G., Liu, L., Meng, Y., et al. Path tracking of wheeled mobile robots based on dynamic prediction model. *IEEE Access*, 7:39690–39701, 2019. DOI: 10.1109/access.2019.2903934. 43

[59] Brown, M., Funke, J., Erlien, S., et al. Safe driving envelopes for path tracking in autonomous vehicles. *Control Engineering Practice*, 61:307–316, 2017. DOI: 10.1016/j.conengprac.2016.04.013. 43

[60] Hwang, C. L., Yang, C. C., and Hung, J. Y. Path tracking of an autonomous ground vehicle with different payloads by hierarchical improved fuzzy dynamic sliding-mode control. *IEEE Transactions on Fuzzy Systems*, 26(2):899–914, 2017. DOI: 10.1109/tfuzz.2017.2698370. 46

[61] Wu, X., Jin, P., Zou, T., et al. Backstepping trajectory tracking based on fuzzy sliding mode control for differential mobile robots. *Journal of Intelligent and Robotic Systems*, 96(1):109–121, 2019. DOI: 10.1007/s10846-019-00980-9. 46

[62] Zhang, C., Hu, J., Qiu, J., et al. A novel fuzzy observer-based steering control approach for path tracking in autonomous vehicles. *IEEE Transactions on Fuzzy Systems*, 27(2):278–290, 2018. DOI: 10.1109/tfuzz.2018.2856187. 47

[63] Taghavifar, H. and Rakheja, S. Path-tracking of autonomous vehicles using a novel adaptive robust exponential-like-sliding-mode fuzzy type-2 neural network controller. *Mechanical Systems and Signal Processing*, 130:41–55, 2019. DOI: 10.1016/j.ymssp.2019.04.060. 47

[64] Xiang, X., Yu, C., Lapierre, L., et al. Survey on fuzzy-logic-based guidance and control of marine surface vehicles and underwater vehicles. *International Journal of Fuzzy Systems*, 20(2):572–586, 2018. DOI: 10.1007/s40815-017-0401-3. 48

[65] Berntorp, K. Path planning and integrated collision avoidance for autonomous vehicles. *American Control Conference (ACC)*, 4023–4028, IEEE, 2017. DOI: 10.23919/acc.2017.7963572. 51

[66] Yuan, Y., Tasik, R., Adhatarao, S. S., et al. RACE: Reinforced cooperative autonomous vehicle collision avoidance. *IEEE Transactions on Vehicular Technology*, 2020. DOI: 10.1109/tvt.2020.2974133. 51

[67] Funke, J., Brown, M., Erlien, S. M., et al. Collision avoidance and stabilization for autonomous vehicles in emergency scenarios. *IEEE Transactions on Control Systems Technology*, 25(4):1204–1216, 2016. DOI: 10.1109/tcst.2016.2599783. 52

[68] Ji, J., Khajepour, A., Melek, W. W., et al. Path planning and tracking for vehicle collision avoidance based on model predictive control with multiconstraints. *IEEE Transactions on Vehicular Technology*, 66(2):952–964, 2016. DOI: 10.1109/tvt.2016.2555853. 52

[69] He, X., Liu, Y., Lv. C., et al. Emergency steering control of autonomous vehicle for collision avoidance and stabilisation. *Vehicle System Dynamics*, 57(8):1163–1187, 2019. DOI: 10.1080/00423114.2018.1537494. 53

[70] Dai, L., Cao, Q., Xia, Y., et al. Distributed MPC for formation of multi-agent systems with collision avoidance and obstacle avoidance. *Journal of the Franklin Institute*, 354(4):2068–2085, 2017. DOI: 10.1016/j.jfranklin.2016.12.021. 53

[71] Dahl, J., de Campos, G. R., Olsson, C., et al. Collision avoidance: A literature review on threat-assessment techniques. *IEEE Transactions on Intelligent Vehicles*, 4(1):101–113, 2018. DOI: 10.1109/tiv.2018.2886682. 55

[72] Schnelle, S., Wang, J., Su, H. J., et al. A personalizable driver steering model capable of predicting driver behaviors in vehicle collision avoidance maneuvers. *IEEE Transactions on Human-Machine Systems*, 47(5):625–635, 2016. DOI: 10.1109/thms.2016.2608930. 56

[73] Kim, J. M., Kim, S. C., Lee, K. H., et al. Preventive effects of seat belts on traumatic brain injury in motor vehicle collisions classified by crash severities and collision directions. *European Journal of Trauma and Emergency Surgery*, pages 1–13, 2019. DOI: 10.1007/s00068-019-01095-4. 57

[74] Vukosavljev, M., Kroeze, Z., Schoellig, A. P., et al. A modular framework for motion planning using safe-by-design motion primitives. *IEEE Transactions on Robotics*, 35(5):1233–1252, 2019. DOI: 10.1109/tro.2019.2923335. 59

[75] Na, X. and Cole, D. J. Game-theoretic modeling of the steering interaction between a human driver and a vehicle collision avoidance controller. *IEEE Transactions on Human-Machine Systems*, 45(1):25–38, 2014. DOI: 10.1109/thms.2014.2363124. 73

[76] Park, M., Lee, S., and Han, W. Development of steering control system for autonomous vehicle using geometry-based path tracking algorithm. *Etri Journal*, 37(3):617–625, 2015. DOI: 10.4218/etrij.15.0114.0123. 73

[77] Samuel, M., Hussein, M., and Mohamad, M. B. A review of some pure-pursuit based path tracking techniques for control of autonomous vehicle. *International Journal of Computer Applications*, 135(1):35–38, 2016. DOI: 10.5120/ijca2016908314. 73

[78] Deur, J., Asgari, J., and Hrovat, D. A 3D brush-type dynamic tire friction model. *Vehicle System Dynamics*, 42(3):133–173, 2004. DOI: 10.1080/00423110412331282887. 75

[79] Ding, N. and Taheri, S. A modified Dugoff tire model for combined-slip forces. *Tire Science and Technology*, 38(3):228–244, 2010. DOI: 10.2346/1.3481696. 75

[80] Guo, H., Chen, H., Ding, H., et al. Vehicle side-slip angle estimation based on Uni-Tire model. *Control Theory and Applications*, 27(9):1131–1139, 2010. 75

[81] Pacejka, H. B. and Bakker, E. The magic formula tyre model. *Vehicle System Dynamics*, 21(S1):1–18, 1992. DOI: 10.1080/00423119208969994. 75

[82] Shao, X., Naghdy, F., and Du, H. Reliable fuzzy H∞ control for active suspension of in-wheel motor driven electric vehicles with dynamic damping. *Mechanical Systems and Signal Processing*, 87:365–383, 2017. DOI: 10.1016/j.ymssp.2016.10.032. 80

[83] Zhang, F., (Ed.) *The Schur Complement and its Applications*. vol. 4, Springer Science and Business Media, 2006. DOI: 10.1007/b105056. 81

[84] Liu, K., Gong, J., Kurt, A., et al. A model predictive-based approach for longitudinal control in autonomous driving with lateral interruptions. *IEEE Intelligent Vehicles Symposium (IV)*, pages 359–364, 2017. DOI: 10.1109/ivs.2017.7995745. 84

[85] NHTSA, Traffic Safety Facts 2017-A Compilation of Motor Vehicle Crash Data (Annual Report), 2019. 91

[86] Jermakian, J. S. Crash avoidance potential of four passenger vehicle technologies. *Accident Analysis and Prevention*, 43(3):732–740, 2011. DOI: 10.1016/j.aap.2010.10.020. 91

[87] Puente Guillen, P. and Gohl, I. Forward collision warning based on a driver model to increase drivers' acceptance. *Traffic Injury Prevention*, 20(sup1):S21–S26, 2019. DOI: 10.1080/15389588.2019.1623397. 91

[88] Yang, W., Zhang, X., Lei, Q., et al. Research on longitudinal active collision avoidance of autonomous emergency braking pedestrian system (AEB-P). *Sensors*, 19(21):4671, 2019. DOI: 10.3390/s19214671. 91

[89] Lee, K. and Peng, H. Evaluation of automotive forward collision warning and collision avoidance algorithms. *Vehicle System Dynamics*, 43(10):735–751, 2005. DOI: 10.1080/00423110412331282850. 93

[90] Tak, S., Park, S., and Yeo, H. Comparison of various spacing policies for longitudinal control of automated vehicles. *Transportation Research Record*, 2561(1):34–44, 2016. DOI: 10.3141/2561-05. 109

[91] Yanakiev, D. and Kanellakopoulos, I. Nonlinear spacing policies for automated heavy-duty vehicles. *IEEE Transactions on Vehicular Technology*, 47(4):1365–1377, 1998. DOI: 10.1109/25.728529. 109

Authors' Biographies

JIE JI

Jie Ji is currently an Associate Professor with the College of Engineering and Technology, Southwest University, Chongqing, China. He received a Ph.D. in Mechanical Engineering from Chongqing University in 2010. From December 2013 to December 2014, he was a Postdoctoral Fellow with the Department of Mechanical and Mechatronics Engineering, University of Waterloo, Ontario, Canada. His active research interests include advanced control & artificial intelligence, and their applications on intelligent and connected vehicles, in which he has contributed more than 30 articles and obtained 9 granted China patents. He is the recipient of the Best Land Transportation Paper Award from IEEE Vehicular Technology Society in 2019, and the Young Professional Excellent Paper Award from China SAE in 2019. Dr. Ji is a member of the China Society of Automotive Engineers, and is the founder of the Green Intelligent Vehicle and Electromobile Laboratory (GIVE Lab) at Southwest University (from 2015).

HONG WANG

Hong Wang is currently a Research Associate Professor at Tsinghua University. From the year 2015–2019, she was working as a Research Associate of Mechanical and Mechatronics Engineering with the University of Waterloo. She received her Ph.D. in Beijing Institute of Technology in China in 2015. Her research focuses on the risk assessment and crash mitigation-based decision making during critical driving scenarios, ethical decision making for autonomous vehicles, component sizing, modelling of hybrid powertrains and intelligent control strategies design for hybrid electric vehicles, and intelligent control theory and application. She becomes the IEEE member since the year 2017. She has published over 50 papers in top international journals, such as *IEEE Transaction on Intelligent System, IEEE Transaction on Vehicular Technology*, etc.

YUE REN

Yue Ren received a B.E. and a Ph.D. in Mechanical and Mechatronics Engineering from Chongqing University, China, in 2013 and 2018, respectively. He is currently working as an assistant professor with the college of engineering and technology, Southwest University, China. He is working on autonomous vehicles, including vehicle detection, path planning and tracking, vehicle dynamics, and stability control.

Printed in the United States
by Baker & Taylor Publisher Services